The Last Steamers

by

Nick Robins

PREFACE

The peak of fast steam reciprocating screw driven ferry design in British and Irish waters were the City of Dublin Steam Packet Company's 'provinces' quartert built in 1896. The ships were capable of 24 knots on the Holyhead to Kingstown mail service, but these fast speeds were soon more economically produced with the turbine engine which first saw commercial service in 1901 on the Clyde. Even though the steam reciprocating engine was no longer preferred for speed, it remained the mainstay of the intermediate ferries well into the twentieth century. The last of these ferries was the *St Clair*, completed in 1937 for the North of Scotland, Orkney and Shetland Steam Navigation Company.

This is the story of the decline of the triple expansion engine. It includes tales of the coastal liner services between London and a variety of British and Irish ports, and of services to the Baltic and Scandinavia and south to Spain and Portugal. It also includes the ubiquitous 'dirty British coaster' and the once plentiful steam tug, all finally driven into submission by the marine diesel engine.

The last passenger steamers finished service in 1969 and the last collier in service finished work for UK owners 1983. The preserved sludge carrier **Shieldhall** carries a lot on her shoulders as the final vestige of the triple expansion engined screw driven coastal steamer.

In telling the story of the last of each class of coastal steamer, a considerable number of people have, of necessity, contributed both information and photographic material. These include Stephen Rabson, P&O archivist; Peter Brooker, Commodore of the Essex Yacht Club; Ian Ramsay, Secretary of the Institution of Engineers and Shipbuilders in Scotland; Malcolm McRonald, acknowledged authority on the history of the Coast Lines Group, and Harold Jordan, master of shipping postcards. The text has also enjoyed thorough editorial review by Ian Ramsay and Malcolm McRonald. Gilbert Mayes has provided much additional information. Assistance with photographic material has been kindly provided, as always, by a number of friends, including Richard Danielson, Donald Meek, Mike Walker, and others. Finally the author is grateful to Bernard McCall at *Coastal Shipping* and Sebright Printers for designing and producing such a handsome product - *The Last Steamers*.

Nick Robins, Crowmarsh, Oxfordshire   December 2005

*A miniature painting of the wreck of the **Berlin** which occurred in February 1907, found in an autograph book dated 1908. The artist was Frank Horne.*

Copyright © 2005 by Nick Robins. The right of Nick Robins to be identified as author of this work has been asserted by him in accordance with the Copyright, Design and Patent Act 1998.

All rights reserved. No part of this publication may be reproduced, stored in a retrieval system or transmitted in any form or by any means (electronic, digital, mechanical, photocopying, recording or otherwise) without prior permission of the publisher.

Published by Bernard McCall, 400 Nore Road, Portishead, Bristol, BS20 8EZ, England.
Website : www.coastalshipping.co.uk
Telephone/fax : 01275 846178. E-mail : bernard@coastalshipping.co.uk
All distribution enquiries should be addressed to the publisher.

Printed by Sebright Printers, 12 - 18 Stokes Croft, Bristol, BS1 3PR
Telephone : 0117 942 5827; fax : 0117 942 0671
E-mail : info@sebright.co.uk
Website : www.sebright.co.uk

ISBN : 1-902953-22-3

CONTENTS

Chapter 1 - The heritage and the competition — 4
Chapter 2 - Fast main-line passenger ferries - Irish Sea — 11
Chapter 3 - Fast main-line passenger ferries - Harwich and South Coast — 19
Chapter 4 - The intermediate ferries — 24
Chapter 5 - Coastal liners — 32
Chapter 6 - Just a few passengers — 39
Chapter 7 - To the islands — 47
Chapter 8 - Of tugs, tenders and excursion ships — 54
Chapter 9 - Dirty British coaster — 62
Chapter 10 - Estuarine steamers — 69
Chapter 11 - Death by diesel — 73
Chapter 12 - In steam again — 77
Appendix — 80
References — 91
More Steam Tugs — 92
More Steam Coasters — 93
Index — 94

*The **Sir Walter Scott**, built in 1900, returns to the Trossachs Pier on Loch Katrine on 21 September 1990.*
(Except where credit is given otherwise, all photographs were taken by the author.)

Cover photographs -
Front cover, top left : The collier **Barford** in the River Thames in November 1969
(World Ship Photo Library, George Gould collection)
Front cover, top right : The **Calshot** attends the **Queen Elizabeth** at Southampton.
(Mick Lindsay collection)
Front cover, lower : Still very much in steam, the **Shieldhall** at Southampton on 13 September 1997.
(Dominic McCall)
Back cover : The **Wallasey** and **Canning** at Swansea on 3 May 1971.
(John Wlitshire)

Chapter 1 : The heritage and the competition

The smell of hot oil, the hiss of steam and the clank of the connecting rods evoke memories of a bygone age. The images of paint peeling from the hot funnel casing on the Promenade Deck and steam leaking from the capstan valve on the forecastle are remembered by a declining number of people as the steamship era becomes just a distant memory. It is easy to think that the marine diesel engine has been in service for all time, but in truth it only began to erode the monopoly of the steam engine in the 1930s and it took more than five decades before steam was defeated.

Today there are just two steam driven coastal passenger ships left in service - the preserved *Waverley* and the former Clyde sludge steamer *Shieldhall*. The *Waverley* is a paddle steamer and the *Shieldhall* is screw driven; they date from 1947 and 1955 respectively. There is also a very remarkable memorial to the screw driven steamer still in operation on Loch Katrine in Scotland, the excursion steamer *Sir Walter Scott*. Built in 1900 with triple expansion engines, her survival today is a tribute to Victorian technology, careful off-season maintenance and her fresh water environment.

The screw driven steam reciprocating engined ship was the stalwart of the majority of passenger carrying routes in the first two-thirds of the twentieth century. But these ships never attracted the glamour of the fast cross-channel steam turbine vessels, nor did they attract the sentiment attached to the paddle steamers which lay at the end of the pier. Despite this, the steam reciprocating engined ships of the UK coasting routes were of great importance to the nation's trade and economy and are worth remembering. Many of these ships were notable for their long and trouble free service or conversely for their colourful careers, but the memory of the men and women who worked aboard these ships or travelled as their passengers is certainly worth recalling.

In the beginning, the very first steam assisted crossing of the Atlantic was that of the paddle wheel driven *Savannah* in 1819. That voyage lasted a traumatic 25 days. By 1838 the passage time had been reduced to a more reasonable $15_{1/2}$ days. Eventually greater speeds became possible with the adoption of the screw propeller and later also the introduction of twin engines and twin screws. Landmarks in the late nineteenth century were the steel hull and the water tube boiler, and by the end of Queen Victoria's reign the steamship was supreme on both the high seas as well as all the short sea services around Britain. Many of the latter still relied on the paddle wheel, but this was to change in the Edwardian era with the introduction of both steam turbine machinery and the development of the triple expansion engine driving a screw propeller, subsequent to which paddle steamers were confined to coastal waters.

*The last British passenger steamship to burn coal was Orkney Steam Navigation Company's **Earl Sigurd** (1931) seen at Kirkwall in 1968.*

The steam turbine was only viable for the faster cross channel passenger traffic and many lesser routes retained the steam reciprocating engine coupled to propeller shafts. The Dover to Calais and Boulogne services of the South Eastern & Chatham Railway and the Stranraer to Larne link of the Midland Railway were unique in passing from a fleet of paddle steamers to one of turbine steamers in the 1900s. Almost all other ferry routes progressed from paddle steamer to propeller driven steam reciprocating engined steamer and only then to turbine or diesel. However, Dr Diesel's newfangled internal combustion technology was yet to be developed and for a long time it seemed that nothing would displace coal burning steam reciprocating engines as the mainstay of the merchant marine.

By way of example the Isle of Man Steam Packet Company commissioned their first screw steamer, the iron-hulled *Mona*, in 1878. Table 1 illustrates the progression within the Manx fleet from wooden hull, through iron to steel. The table also shows the introduction of the oscillating engine in 1863 and the demise of the side-lever engine as well as the increase in size and operating speed of vessels in the late nineteenth century. The compound engine linked to propellers was first introduced to the fleet in 1878 and the triple expansion engine in 1891. More significantly the steam turbine engine first entered Douglas Harbour in 1905.

The *Mona* was a single screw vessel with inverted compound engines. The *Fenella* followed only three years later, in 1881, complete with twin screws and compound engines; the working steam pressure was a paltry 85 psi. In 1882 the 1860-built *Mona's Isle* was converted from a side-lever engined paddle steamer to a compound engined twin screw steamer and renamed *Ellan Vannin*; the company was obviously pleased with the performance of the *Mona* and *Fenella*. Further paddle steamers were built for the company, however, and the next ship was the steel-hulled paddler *Mona's Isle,* which was commissioned in 1882.

Four more big and fast Manx paddle steamers were built. One more compound engined screw steamer was the *Peveril*, a replacement for the first screw steamer the *Mona* which had been lost in 1883. Two triple expansion screw steamers, the *Tynwald* built in 1891 and the *Snaefell*, in 1910, were both designed for all year round service and specifically for winter duty. By 1905 the company had taken delivery of its first direct-drive turbine steamer, the *Viking*. Although steam reciprocating engined tonnage was subsequently bought on the secondhand market, no more were built for the company's main line passenger services after the *Snaefell*, and all subsequent new-build tonnage for the main line passenger routes was steam turbine and later diesel.

*The **Lyonesse** (1889) operated the Scilly Isles service for the West Cornwall Steamship Company until 1918. She was one of the first coastal steamers to be equipped with a triple expansion engine.*

(Author's collection)

Notwithstanding, steamers were yet to be built for a variety of other owners both for the coastal and short sea routes and the longer international passenger services. Steam reciprocating engines remained supreme in coastal and international dry and liquid cargo trades until well after World War Two. However, by 1960 there were only three British registered (triple expansion) steam reciprocating engined passenger liners left on the register. These were the British India Line's venerable stalwart, the *Rajula*, dating from 1926 and the former Natal Line vessels *Umtali* and *Umtata* which ended their days under Elder Dempster Line colours as the *Calabar* and *Winneba* respectively.

*Last of many liners, British India Line's steamship **Rajula** (1926) continued in service between Singapore and Madras until finally being withdrawn from UK registry in 1973.*

(P&O Group)

In 1960, there were also precious few passenger steamers left on the short sea services. The North of Scotland, Orkney and Shetland Shipping Company was still operating the **St Clair** and **St Magnus** dating from 1937 and 1924 respectively. The City of Cork Steam Packet Company had the 1936-built **Glengariff** in passenger service between Cork and Liverpool, and the British Transport Commission maintained the **Great Western** in passenger service between Fishguard and Waterford. However, although the former cargo and passenger coastal liner **Ulster Herdsman** was still operating for the Belfast Steamship Company, she had not carried passengers for many years. There were also the combination triple expansion and low pressure turbine engined ships of the Ellerman's Wilson Line that were still in operation.

*The **Borodino** accommodated 56 passengers on Wilson Line's Hull to Copenhagen service until withdrawn in 1971. She was one of the last steamers to be built with the combination triple expansion and low pressure turbine system.*

(Author's collection)

Apart from the remaining paddle steamers in service on the Clyde, Bristol Channel, South Coast and Thames, the only opportunity to travel on a steam driven screw steamer in the 1960s was on the inter-Orkney services, the Plymouth tenders, some of the Mersey ferries, the North to South Shields or the Tilbury to Gravesend vehicle ferries. There were also a few small estuarine excursion steamers then also in service. The Atlantic Steam Navigation Company had several former Landing Craft Tank which could accommodate twelve passengers still running out of Tilbury and Preston, but passengers were mainly carried on the new diesel ferries **Bardic Ferry** and **Ionic Ferry**. At Harwich the Zeeland Steamship Company had withdrawn the venerable Dutch-flagged **Mecklenburg** from relief duties from the Hook of Holland only in October 1959.

Of course, there remained a wide variety of coastal cargo vessels, tugs and other craft that were dependent on the steam reciprocating engine. The last of the estuarial paddle steamers had been commissioned in the late 1940s. There was a number of steam turbine ferries yet to be built for the railways and the Isle of Man Steam Packet Company during the 1960s, but the heyday of the steamer was long past. Their death knell was rising fuel costs, the increasing efficiency of the motor ship and an increasing number of experienced motor ship engineers.

The stalwarts of the early twentieth century on the coastal services were very much the steam reciprocating screw driven steamers. The main benefits of steam machinery were availability and low costs of coal or fuel oil, an abundance of steam certificated engineers, and relatively simple and easily maintained technology. The main drawback was that the weight of the engines and boilers full of water, and the size of the engine spaces required for a steam reciprocating engine, far exceeded those of a diesel engine so reducing the relative payload of the steamer. The other attraction of the diesel was that engines can be shut down and started up at ease without having to prepare the boiler and get up steam.

*The **Amsterdam** (1894) of the Great Eastern Railway's Harwich fleet exemplified the classic appearance of the late-Victorian cross-channel steamer.*

(Author's collection)

*Pinnacle of design of fast triple expansion engined ferries was the **Hibernia** (1900), seen leaving Dublin. With her three sisters she maintained the service to Holyhead for the London & North Eastern railway.*

(Mike Walker collection)

After the successful introduction of the *King Edward* on Clyde services, high speed turbine engined vessels were introduced in the English Channel and the Irish Sea from 1903 and 1904 respectively. The fast and relatively efficient turbine steamers dominated the cross channel passenger trade for the next sixty years, only being ousted by escalating fuel oil costs in the 1960s [1].

The least successful of the early direct-drive turbine steamers were the *Immingham* and *Marylebone*, built in 1906 by Swan, Hunter & Wigham Richardson on the Tyne and Cammell Laird at Birkenhead respectively for the Great Central Railway's nightly service between Grimsby and Rotterdam. Their design speed was 18 knots, modest for the new turbine steamers, but reducing the voyage time from 17 hours to 10. They were expensive units to operate and, with competition for passengers from the Great Eastern Railway at Harwich, they were soon re-engined and down-graded to triple expansion steam reciprocating engined ships, with a boiler pressure of 160 psi. Thus from 1911 onwards the pair could only maintain 13 knots but were earning their keep until the intervention of the Great War. The service was not resumed thereafter. Without exception the other 126 turbine steamers in service in British waters were a huge success.

[1] See *Turbine Steamers of the British Isles, 1999.*

The first motor ferries were the *Ulster Monarch* and her various sisters on the Irish Sea, and the smaller units of David MacBrayne's West Highland fleet which were commissioned from 1930 onwards [2]; all subsidiaries of the Coast Lines group. The diesel took a long time to eat into the steam reciprocating engine's monopoly on the slower overnight passenger services and the coastal cargo and passenger liner routes. Coastal cargo ships were largely steam driven until after World War II with the exception of the diesel fleets of Coast Lines and a few other far seeing operators.

Although paddle steamers had been ousted from the main line passenger routes early in the twentieth century, they remained in vogue in estuarial waters until well into the 1950s. All were powered by steam reciprocating machinery save for a handful of diesel electric and diesel-hydraulic vessels, notably the *Talisman* in the North British Railway fleet, the Forth vehicle ferries and the Isle of Wight ferry *Farringford*.

This story concentrates on the steam reciprocating screw driven steamers of the British coasting trades in the twentieth century. It does not deal with the story of the paddle steamer, neither does the book attempt to chronicle all the UK registered steam vessels involved in passenger and cargo coastal liner routes, but rather highlights the last groups of each type of ship as well as the more significant vessels on the more important routes. It is intended as a tribute to their crews and passengers alike in the hope that memories of the steam era will be kept alive for a generation that never had the pleasure of experiencing the ambience of the steamship.

*The starboard triple expansion engine in the workshop at Vickers yard in Barrow, ready for installing in the **Duke of Lancaster** (1895), built for the Lancashire & Yorkshire and London & North Western Railway's service between Fleetwood and Belfast.*

(Barrow Museum Service)

[2] See *Ferry Powerful, a History of the Modern British Diesel Ferry*, 2003.

Steam Reciprocating Engines

The basic principle of the steam engine is a cylinder in which sits a piston connected to a piston rod. The piston rod passes through one end of the cylinder through a stuffing box and is connected to the connecting rod which in turn connects with the crankshaft (Figure 1). The cylinder is vertical so that the crankshaft lines up with the propeller via the thrust shaft, tunnel shaft and tail shaft. The engine is called a reciprocating engine as the piston rises and falls or reciprocates back and forth within the cylinder. Superheated steam drawn via a throttle valve from the boiler passes into the cylinder via a slide valve. The steam trapped in the cylinder expands and forces the piston to move away from the top or bottom of the cylinder and this motion allows the slide valve to cut off the steam. As the piston nears the far end of the cylinder, the slide valve allows the spent steam to evacuate the cylinder and new steam under pressure from the boiler to enter the cylinder at the opposite end, again expanding to push the piston back again. The slide valve is driven by an eccentric rod connected to an eccentric disk or sheave on the crankshaft so that the slide valve is consistently out of phase with the piston. Unlike a petrol or diesel engine in a car, which is single acting, the steam engine is double acting.

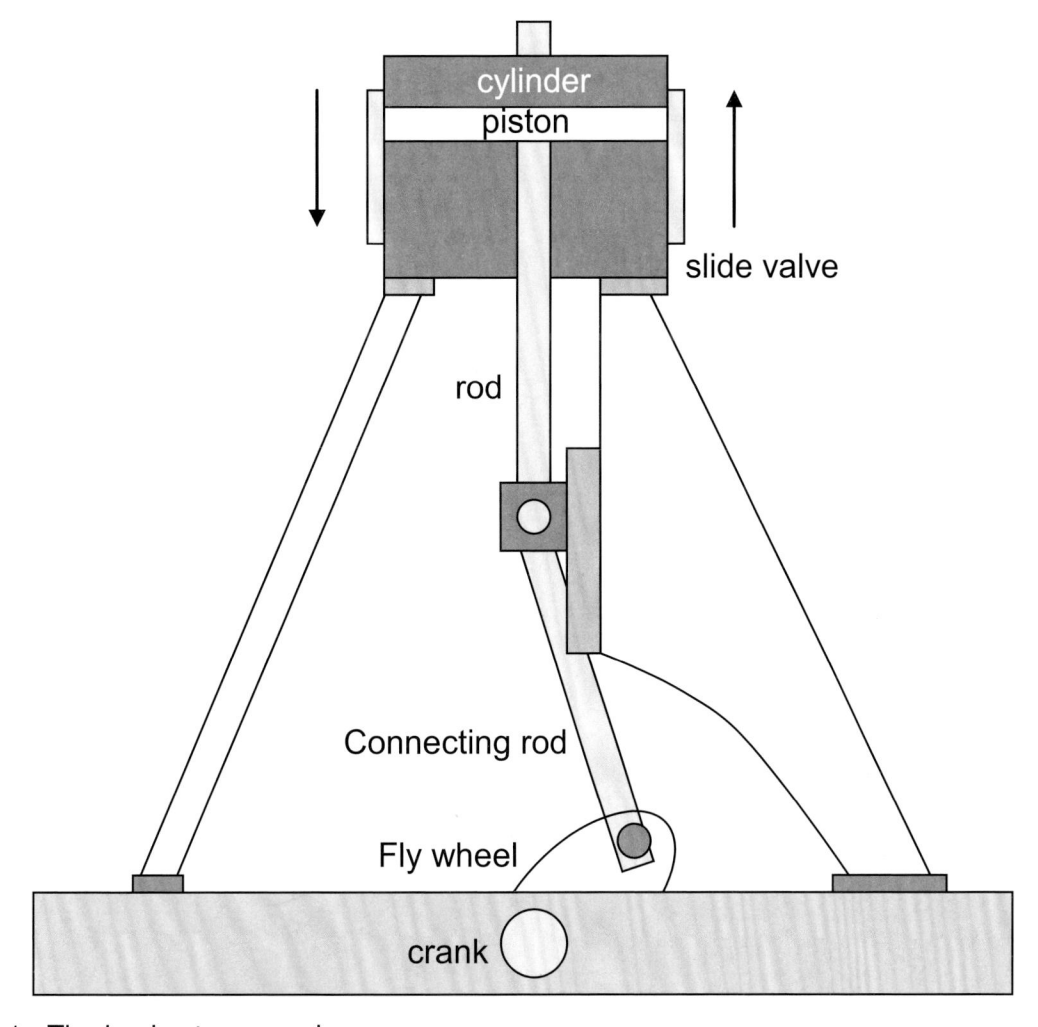

Figure 1 : The basic steam engine

The spent steam can be passed into a larger cylinder as it is now at lower pressure, and the process repeated to extract more energy from the steam. This is the compound steam engine. A third even larger cylinder and a fourth comprise the triple and quadruple expansion engine. Some triple expansion engines had two parallel low pressure cylinders of equal diameter and receiving steam in parallel. Alternatively the triple expansion engine can exhaust via a low pressure turbine to provide a composite engine, a system that was very popular with the Ellerman's Wilson Line, utilising the Baur Wach Exhaust Turbine. The steam is returned to the boiler via the condenser which returns the remaining steam back to a condensate by bringing it into contact with cold pipes. The pipes are cooled by sea water driven through them by the circulating pump. The condensate is sucked out of the condenser by the air pump which maintains a vacuum in the condenser. The condensate is returned to the boiler via the hot well tank and its pump and the feed heater and the feed pump. In this way the loss of feed water is minimised.

TABLE 1 : Early steamers built for the Isle of Man Steam Packet Company.

Ship	Launch date	Tons gross	Hull	Drive	Engines	Speed (knots)
Mona's Isle	1830	200	Wood	Paddle	Side lever	8.5
Mona	1832	150	Wood	Paddle	Side lever	9.0
Queen of the Isle	1834	350	Wood	Paddle	Side lever	9.5
King Orry	1842	433	Wood	Paddle	Side lever	9.5
Ben-my-Chree	1845	399	Iron	Paddle	Side lever	10.0
Tynwald	1846	700	Iron	Paddle	Side lever	14.0
Mona's Queen	1852	400	Iron	Paddle	Side lever	13.0
Douglas	1858	700	Iron	Paddle	Side lever	15.0
Mona's Isle	1860	380	Iron	Paddle	Side lever	12.0
1882 - renamed *Ellan Vannin*		342		**Twin Screw**	**Compound**	12.5
Snaefell	1863	604	Iron	Paddle	Oscillating	15.0
Douglas	1864	709	Iron	Paddle	Oscillating	15.0
Tynwald	1866	700	Iron	Paddle	Oscillating	15.0
King Orry Lengthened 1888	1871	806	Iron	Paddle	Oscillating	14.0
Ben-my-Chree	1875	1030	Iron	Paddle	Oscillating	14.0
Snaefell	1876	849	Iron	Paddle	Oscillating	15.0
Mona	1878	526	Iron	**Single screw**	**Compound**	13.5
Fenella	1881	564	Iron	**Twin screw**	**Compound**	13.5
Mona's Isle	1882	1564	Steel	Paddle	Compound oscillating	17.5
Peveril	1884	561	Steel	**Twin screw**	**Compound**	13.5
Mona's Queen	1885	1559	Steel	Paddle	Compound oscillating	18.0
Queen Victoria	1887	1657	Steel	Paddle	Compound diagonal	20.5
Prince of Wales	1887	1657	Steel	Paddle	Compound diagonal	20.5
Tynwald	1891	937	Steel	**Twin screw**	**Triple expansion**	18.0
Empress Queen	1897	2140	Steel	Paddle	Compound diagonal	21.5
Viking	1905	1957	Steel	Triple screw	Direct-drive turbine	23.0
Ben-my-Chree	1908	2651	Steel	Triple screw	Direct-drive turbine	24.5
Snaefell	1910	1368	Steel	**Twin screw**	**Triple expansion**	19.0
King Orry	1913	1877	Steel	Twin screw	Geared turbine	21.5

Screw-driven steamers with reciprocating engines have **drive** and **engine** details in bold type.

Chapter 2 : Fast main-line passenger ferries – Irish Sea

During the 1900s the steam turbine engine became the preferred power source for most operators who required fast day time crossings. The railway companies in particular turned their backs on the triple expansion engine in favour of the turbine. But the steam reciprocating engine was almost universally retained for the slower overnight services and was still retained for some of the faster day services as well. One operator, the Coast Lines Seaway group of companies, in due course went from the triple expansion engine to the diesel engine in one step with the introduction of the *Ulster Monarch* in 1929.

In the early 1900s, when William Denny and Brothers and their sub-contractors were putting the finishing touches to the *Sir Walter Scott* on Loch Katrine, the triple expansion engined screw steamer was at its zenith. The design and construction of the faster steamers peaked in the 1900s as the more efficient direct-drive turbine steamers came on line. State of the art at the turn of the century were the quartet of fast steamers commissioned by the City of Dublin Steam Packet Company in 1896 for the Kingstown to Holyhead mail service. This prestigious service provided the important contact between London and the regional administration in Dublin. The four new ships replaced an older quartet which had carried the same names, and which had been built in 1860, these were the *Ulster*, *Munster*, *Leinster* and *Connaught*, being the four provinces of Ireland.

The four new ships were large, with a gross tonnage of 2300, and they were fast. They were capable of an incredible 24 knots and could make the crossing, pier to pier, in just $2^{3}/_{4}$ hours. When the last of the ships was delivered in 1897 they were the largest and fastest cross-channel steamers in the world. They had four double-ended boilers which delivered steam at 175 psi to a pair of four-cylinder triple expansion engines (twin low pressure cylinders) coupled to twin screws. As it turned out, these were the fastest ever steam reciprocating engined cross channel ferries.

The London & North Western Railway, which maintained the passenger service between Dublin North Wall (Kingstown from 1908 onwards) and Holyhead, responded to the challenge of the *Ulster* and her sisters with a new quartet of its own. These were the *Cambria*, *Hibernia*, *Anglia* and *Scotia*, which were completed at William Denny's yard at Dumbarton between 1897 and 1902. First class accommodation was on the Main Deck and saloon class aft on the Main Deck and also on the Lower Deck. The ships were smaller than the *Ulster* and her sisters with a gross tonnage of 1800, but featured the same arrangement of twin funnels each topped by cowls.

*The **Anglia** leaves the Inner Harbour at Holyhead with the hotel in the background.*

(Mike Walker collection)

The *Cambria* and her sisters were equipped with a pair of fast running four-cylinder triple expansion engines which were fed by coal-fired Scotch boilers to give the ships a service speed of 21 knots. The cylinders were 26, 40 and two by 43 inches and the stroke was 33 inches. They were designed with a slower speed than the City of Dublin Steam Packet Company's mail ships as these ships did not need to compete and also had the slow run up along the Great South Wall into the River Liffey to contend with. The faster speed of the City of Dublin vessels was a requirement of the mail contract with penalties for late arrivals. Nevertheless, the London & North Western Railway's ships were the second fastest group of British steam reciprocating engined cross channel ferries.

Ironically, the City of Dublin Steam Packet Company lost the mail contract to the rival London & North Western Railway in 1920. The surviving pair of mail ships, the *Ulster* and *Munster* (the *Leinster* and *Connaught* having been lost in the Great War, the former with considerable loss of life) completed their last crossings on 27 November and were laid up pending sale to the breakers yard. The *Anglia* and *Hibernia* had also been lost in the war and the *Cambria* and *Scotia*, temporally renamed *Arvonia* and *Menevia* towards the end of their careers, were replaced for the new mail contract by another quartet, this time fast turbine ships, but with the same kingdom names.

Holyhead was also home to that part of the London to Belfast service operated by the London & North Western Railway via Greenore, a small port on Carlingford Lough south of Newry. The service was opened in 1873 in direct competition with the Midland Railway at Barrow and Morecambe. The Irish link included 12 miles of new railway line from Greenore to Newry and purchase of the Dundalk & Greenore Railway. Initiated with a fleet of paddle steamers, the first triple expansion engined screw steamers built for the route were the Denny-built *Rosstrevor* and sister *Connemara*. They had twin engines each with cylinders of 19, 29 and 44 inches with a stroke of 30 inches. A third ship completed the trio to displace the last of the paddlers in 1898. This was the *Galtee More*, but she had four-cylinder triple expansion engines with cylinders of 19, 29, and two at $31\frac{1}{2}$ inches, again with a stroke of 30 inches.

The four-cylinder arrangement was repeated ten years later when the next generation of steamer arrived on the service. This was the *Rathmore*, completed in 1908 with twin engines each with 25, 37 and two by 41 inches with the same 30 inch stroke. This larger cylinder set provided an increased service speed from 18 knots on the original trio to 20 knots. She was followed by two turbine steamers, one delivered before the Great War and one after it, which could manage an extra half knot over the *Rathmore*. At the outbreak of hostilities the new turbine driven ship was transferred to the Holyhead to Kingstown service as the *Anglia* and her sisters were requisitioned by the Admiralty. The original trio of steamers plodded on through the war on their own route.

*The **Connemara** seen on trials.*

(Mike Walker collection)

The hazards of the war not only included the enemy but also the need to traverse the Irish Sea without navigation lights and in total darkness. A C Yeates described the fatal end of the *Connemara* in an article which first appeared in *Sea Breezes* in January 1968:

On Friday night, 3 November 1916, the *Connemara* with Captain G H Doeg in command, left Greenore for Holyhead in the teeth of a strong south westerly gale and was due at Holyhead in the early hours of Saturday morning. The *Galtee More* left Holyhead early on Saturday morning and passed what they assumed to be the *Connemara* shortly after leaving the harbour. As the *Galtee More* was entering Carlingford Lough in the early light of Saturday morning, her master sighted what appeared to be a stranded submarine. This he reported immediately upon his arrival and it was not until then that he learned that the *Connemara* had not reached Holyhead. A party immediately left to investigate and found that the supposed submarine was the wreck of the *Connemara* lying on her side, and not far from her was the wreck of another vessel, later identified as a small collier.

The sole survivor from the two ships recounted that the collier had developed a list as her cargo shifted and had become unmanageable. Hit by the *Connemara* at speed, the impact was such that no time was available to launch lifeboats before both ships sank. All 82 passengers and crew aboard the *Connemara* and eleven members of the collier's crew lost their lives that night, and all because the ships were blacked out. Later on in the war, the *Rosstrevor* was the subject of a submarine attack, but the torpedo ran safely to one side of the ship as she approached Holyhead.

Although the Greenore service survived the war it did not survive the changing post-war scene. In order to boost income, the ships were allowed to carry out occasional excursions and charters; for example, the *Rathmore* made at least one full day excursion from Barrow to Llandudno. But with the Partitioning of Ireland in 1922, and the Dail confirming the Irish Treaty, passengers to Belfast via Greenore now had an international border to cope with outside Newry. The end of the service was inevitable, with depressed traffic conditions and the rationalisation of the railway companies that followed in 1923. The service closed in 1926. The new London, Midland and Scottish Railway had taken over the London and North Western and the Midland railways and could see little benefit in keeping the Greenore service going in competition with its own direct Heysham to Belfast service. Whilst the surviving *Rosstrevor* and *Galtee More* were scrapped, the *Rathmore* served a further five years as the *Lorrain* on a new Tilbury to Dunkerque route for her owners in an Anglo-French consortium.

There were nearly forty other ships capable of over 18 knots on the British register (Table 2). None could compete with the crack Holyhead vessels. The Fleetwood Dukes belonging to the Lancashire & Yorkshire and London & North Western railways could also maintain 20 knots on services to Belfast and Douglas. The first of the twin screw triple expansion engined ships to succeed the paddle steamers was the *Duke of Clarence* built by Laird of Birkenhead in 1892. The second of the class was the *Duke of York*, and she went on to become the Isle of Man Steam Packet Company's *Peel Castle* in 1912, eventually being sold for scrap in 1939 at the age of 45. The third ship was built at Barrow as the *Duke of Lancaster* and she too went to the Isle of Man Steam Packet Company in 1912 as *The Ramsey*, alas lost in war three years later. The next ship came from Vickers, the *Duke of Cornwall*, which also became an Isle of Man boat as the *Rushen Castle* when she was sold in 1928. She survived the second war to be sold for demolition in 1947. Dr David McNeill [3] (see references) described the *Duke of Cornwall*:

She was very similar to the other four vessels in her class with a steel hull and a single funnel some 35 feet (11 metres) high. Her saloon passengers were berthed amidships and her steerage aft. The sleeping accommodation for men in the latter was reserved for cattle drovers; other male passengers were considered to be tough and could sleep either on deck or on the wooden benches in the steerage covered accommodation. Saloon passengers had two and four berth cabins. Their dining saloon was on the Main Deck and had a handsome fireplace with an open fire. Ladies travelling on their own, or in all-female parties, were accommodated in either a dormitory or special four berth cabins situated on the port side of the Main Deck. These were ruled by a benevolent dragon in the form of a buxom stewardess in starched white apron and bonnet to match. On the deck above was a well furnished and adequately stocked smoke room which in those days was patronised by gentlemen only.

*The **Duke of Cornwall** (1898) seen on trials. She was a slightly modified sister of the **Duke of Lancaster** (1895) whose engines are shown on page 8.*

(Mike Walker collection)

[3] David McNeill was both Physics lecturer and hall tutor to the author in the late 1960s at the University of Southampton. He later retired to his native Belfast.

The final pair of steam reciprocating engined ships were the *Duke of Connaught* and *Duke of Albany* which came from John Brown on Clydebank. The *Duke of Connaught* was given new Babcock & Wilcox water tube boilers after the Great War following a brief stint on the mail contract on charter to the City of Dublin Steam Packet Company. The very last Fleetwood based Dukes, however, were the Denny-built sisters *Duke of Cumberland* and *Duke of Argyll*, but these were equipped with steam turbine engines. In 1922 the Lancashire & Yorkshire and London & North Western railways amalgamated and within a year had become part of the London, Midland & Scottish Railway. The Fleetwood services were closed in favour of Heysham in 1928 by the London, Midland & Scottish Railway and the vessels dispersed to other routes.

Since the 1860s services from Barrow and Morecambe to Belfast had been maintained by the Barrow Steam Navigation Company which was partly owned by the Midland Railway. In response to the Fleetwood Dukes the company ordered four near sisters, two with triple expansion engines and two with turbine engines. Each was designed by Sir John Biles, and the new ships started between the newly built port of Heysham and Belfast in 1904. The four-cylinder triple expansion engined *Antrim* and *Donegal* were built respectively by John Brown and Caird and the direct-drive, triple screw turbine ships *Londonderry* and *Manxman*, were respectively built by Denny and Vickers. Each had very high quality first and saloon class accommodation. The *Manxman* on the summer-only day service to Douglas was licensed to carry over 2000 passengers. The *Manxman* relieved on the Belfast service during winter refits. Captain Isherwood reported in *Sea Breezes*, May 1972:

The comparison between the reciprocating and turbine-engined ships was especially interesting, the hulls being the same, and it was found that the *Londonderry* had about one knot more speed on a lower fuel consumption than the *Antrim* and *Donegal*, while the *Manxman* was about ³/₄ of a knot faster than the *Londonderry*.

The *Donegal* was a war victim being lost in 1917. The *Antrim* continued on service until displaced by a new generation of turbine steamers in 1928 when she was sold for ten years further service with the Isle of Man Steam Packet Company as their *Ramsey Town*.

The Antrim (1904) at Heysham. This view is taken from a postcard which says on the reverse "affix halfpenny stamp".

(Author's collection)

The Belfast Steamship Company built a series of distinguished steamers for the Liverpool service. The first of note was the *Magic* which was introduced to the service in 1893 with a reported trials speed of 19 knots, followed by the *Graphic* and *Heroic* of 1906 and finally the *Patriotic* which was only completed in 1911. Exceptionally, the *Graphic* and *Heroic* had quadruple expansion engines rather than the more common triple arrangement. Until the *Graphic* and *Heroic* arrived on the scene, the running mates of the *Magic* were only capable of 14 knots so her eight hour passage was sought after by passengers who did not relish the thought of the alternative eleven hour transit on the older vessels. Financial problems postponed the order for the second and third upgraded vessels for 13 years.

Sinclair, in his book *Across the Irish Sea*, described the ***Magic***:

The saloon was handsomely decorated in polished oak with tastefully carved and ... costumed figures. The deck was laid with parquetry flooring, and the whole was surmounted by an elegant dome with stained glass. The sleeping accommodation was arranged on the Saloon, Main and Lower decks and was for 222 passengers. No stateroom accommodated more than four passengers; all were mechanically ventilated and equipped with electric bells. In the more spartan steerage, separate sleeping accommodation was provided for women. An innovation here was a supplementary charge of one shilling for which a women could have a berth with tea or coffee in the morning included.

Ventilation was none too good in those days and inside cabins could be hot and sticky on a calm summer night. In addition the ***Magic*** was afflicted with a vibration problem. When the company could finally manage to order new express tonnage for the Liverpool service there was much agonising over the machinery. Direct-drive turbines were considered for the new pair of ships. However, the deeper draught of a high speed reciprocating engined vessel was not a problem and the company was assured by the builders, Harland &Wolff, that the balanced quadruple expansion engine, trialled by the White Star Line's ***Celtic*** of 1901, would guarantee vibration free operation. Mechanically the outcome was a success, with a unique pair of ships each equipped with a pair of quadruple expansion engines and claimed to be a little faster than the ***Magic***; the ***Graphic*** was always a little faster than her sister the ***Heroic***. Hindsight must, however, question why these ships had not been equipped with turbines, which, by the time they were built, had already demonstrated smoothness of operation and relative efficiency. Nevertheless, the new ships put the service back into serious competition with the Midland Railway at Fleetwood and helped place the Belfast company back onto a better financial footing.

The last and finest of the four ships was undoubtedly the ***Patriotic***. Capable of 18 knots, she was finely appointed and equipped with the more conventional triple expansion engines. Completed in 1911, she was also the very last of the fast steam reciprocating ferries to be built for the British register. On departure from York Dock in Belfast for trials she saluted the new White Star liner ***Titanic***, and indeed had that company's marine superintendent aboard as a guest.

All four ships led long lives. As a hospital ship in the Great War, the ***Magic*** became confused with a new destroyer of the same name and adopted the suffix *II*, later adopting the name ***Classic***. The post-war service was resumed by the four ships minus the ***Heroic*** which was not released from military duty until July 1920. In 1923 the ***Graphic*** was sunk in Belfast Lough by an American steamer. There was no loss of life and she was raised, refurbished and later put back into service. All four ships were converted from coal to oil burning in 1924, when the ***Classic*** transferred to the City of Cork Steam Packet Company at a price of £40 000 to become the ***Killarney***. At the start of the Great Depression she became a seasonal cruise 'yacht' being laid up in Liverpool throughout the winter.

*The first of the 19-knot quartet for the Belfast Steamship Company was the **Magic** (1893). Renamed **Classic** in the Great war and later becoming **Killarney**, she is seen in cruising livery in a West Highland setting.*

(Author's collection)

The other three vessels were displaced from the Liverpool service with the arrival of the new diesel ferry *Ulster Monarch* and her sisters. The steamers were transferred to the Liverpool to Dublin route of the British & Irish Steam Packet Company. The *Graphic* was renamed *Lady Munster* in June 1929, and later *Louth*, and in March 1930 the *Heroic* became *Lady Connaught*, and later *Longford* and the *Patriotic* became *Lady Leinster*, confusingly renamed again as *Lady Connaught* in 1938. Each had its funnel lowered and a second dummy funnel installed.

All survived the Second World War. The *Lady Connaught* (ex *Patriotic*) was mined in December 1940 and put back to Liverpool. During repairs at Dublin her entire boat deck was removed and with it all the single berth cabins on the deck below. These were replaced on her conversion to a hospital ship in 1943 when she became home to American medical staff.

The former *Magic* was sold after the war to Greek owners later being wrecked in 1951. The *Louth* and *Longford* reopened the Belfast to Liverpool service in 1945 before transferring to the Dublin to Liverpool service. Displaced by new tonnage, the *Louth* received a special survey and was renamed *Ulster Duke* in 1948.

The former *Patriotic* ended her days as the cruise ship *Lady Killarney*, popular on short seasonal cruises to Scotland until withdrawn and scrapped in 1956. The old *Graphic*, as *Ulster Duke*, was no longer required on the Belfast route as access to Princes Dock, Liverpool, had then been made independent of tides requiring only two ships, and she was sold in 1951. However, she foundered whilst under tow to her new Italian owners. The *Heroic*, as *Longford*, remained as relief ship on the Irish Sea until sold for demolition in 1952, displaced by the new diesel engined relief ship *Irish Coast*. A truly remarkable record for any group of cross channel ships; the total years in service of the four steamers amounted to 194 years.

From the commissioning of the *Patriotic* in 1911, steam reciprocating engines would never again be designed for high speeds. Notwithstanding, the reciprocating engine remained the mainstay for UK ferry and short sea passenger and cargo operators for a long time to come.

*Laird Line's **Hazel** (1907) seen at Portrush. She was one of a number of secondhand purchases made by the Isle of Man Steam Packet Company after the Great War when she became its **Mona** until sold for demolition in 1938.*

(P&O Group)

The sinking of the *Leinster* from an article by Jill Tunstall and Alun Prichard, which first appeared in the *Daily Post* (North Wales) 22 September 2003.

There would have been mixed emotions on the quayside as the mail boat *Leinster* slipped its moorings on the grey morning of 10 October 1918 and headed out from Dun Laoghaire for Holyhead. On board were soldiers and civilians. These were the dying days of the Great War, when disasters at sea had become daily headlines with civilians and soldiers alike falling prey to U-boats.

Understandably those stepping aboard would have been nervous of the prospect of attack. But there was optimism too for the British, Canadian and Australian soldiers were soon to go home after years fighting on the European Fronts and the suppression of the Easter Rising in Ireland.

Shortly before 10am, Oberleutnant Robert Ramm, the 27 year old captain of the German submarine *UB-123*, found the *Leinster* in his sights. It is not known whether he knew that there was civilians and children on board but, carrying out his duty, he gave the order to fire the first of three torpedoes. It passed across the bow, but the second struck the port side in the vicinity of the ship's mail room. Only one of the postal workers survived the sinking. There was panic as the 700 passengers and 77 crew were rocked by the blast. 501 passengers were killed that day.

In the wake of the second blast Captain William Birch struggled to turn his ship round to head back to Dun Laoghaire. His German counterpart had different ideas and gave the order for a final torpedo, which struck the *Leinster* on the starboard side.

There had been three earlier attacks on the mail boats and the *Ulster* had only just managed to zig-zag to avoid a torpedo earlier in the day. Ironically, and despite requests for escorts by the mail boats at the time, the British Admiralty claimed they would slow them down and make them more of a target.

Apart from postmen and civilians, the mail boats carried thousands of troops during the war, to mobilise reinforcements to help suppress the Easter Rising in Dublin. Censorship and sensitivity about conscription and home rule in Ireland meant that details of the sinking were suppressed, and the Stop Press edition of one Dublin paper was seized and withdrawn. The Foreign Secretary, Arthur Balfour, said the *Leinster* was carrying no military stores and serving no military object.

Of the military men aboard, 67 were officers and 425 other ranks. Only 164 survived. Roy Stokes, author of the book *Death in the Irish Seas: the Sinking of the RMS*, says:

"The sinking was a major loss, but there is a whole untold story about a huge U-boat war in the Irish Sea in 1917-18. Hundreds of ships were sunk and many sailors were lost. It is part of our history we just don't know anything about, despite the fact it happened on our doorstep."

Very little was done about the sinking by a war weary public, and just a month later the conflict ended. Oberleutnant Robert Ramm, however, died before the ceasefire when his boat sank after hitting a mine on its way back to Germany.

Table 2 The fastest main line reciprocating engined steamers

Ship	Built	Owners	Tons Gross	Speed (knots)	NHP
Connaught	1897	City of Dublin Steam Packet Company	2632	24	529
Leinster	1897	City of Dublin Steam Packet Company	2632	24	529
Munster	1897	City of Dublin Steam Packet Company	2632	24	529
Ulster	1897	City of Dublin Steam Packet Company	2632	24	529
Anglia	1900	London & North Western Railway	1872	21	425
Cambria	1897	London & North Western Railway	1842	21	425
Hibernia	1900	London & North Western Railway	1862	21	424
Scotia	1902	London & North Western Railway	1872	21	425
Rathmore	1908	London & North Western Railway	1569	20	742
Duke of Clarence	1892	Lancashire & Yorkshire and London & North Western Railway	1458	20	500
Duke of York	1894	Lancashire & Yorkshire and London & North Western Railway	1531	20	457
Duke of Lancaster	1895	Lancashire & Yorkshire and London & North Western Railway	1520	20	350
Duke of Cornwall	1898	Lancashire & Yorkshire and London & North Western Railway	1540	20	340
Duke of Connaught	1902	Lancashire & Yorkshire and London & North Western Railway	1680	20	340
Duke of Albany	1907	Lancashire & Yorkshire and London & North Western Railway	2259	20	425
Antrim	1904	Midland Railway	2100	20	386
Donegal	1904	Midland Railway	1997	20	386
Seaford	1894	London, Brighton & South Coast Railway	997	20	292
Sussex	1896	London, Brighton & South Coast Railway	1117	20	292
Arundel	1900	London, Brighton & South Coast Railway	1067	20	292
Fredericia	1890	London & South Western Railway	1059	19	494
Lydia	1890	London & South Western Railway	1059	19	494
Stella	1890	London & South Western Railway	1059	19	494
Alma	1894	London & South Western Railway	1145	19	414
Columbia	1894	London & South Western Railway	1145	19	414
Alberta	1900	London & South Western Railway	1240	19	485
Reindeer	1897	Great Western Railway	1281	19	643
Roebuck	1897	Great Western Railway	1281	19	643
Magic	1893	Belfast Steamship Company	1630	19	493
Hazel	1907	Laird Line	1241	19	463
Snaefell	1910	Isle of Man Steam Packet Company	1368	19	
Galtee More	1898	London & North Western Railway	1112	19	430
Ibex	1891	Great Western Railway	1150	18	560
Rosstrevor	1895	London & North Western Railway	1094	18	420
Connemara	1897	London & North Western Railway	1105	18	420
Graphic	1906	Belfast Steamship Company	1871	18	788
Heroic	1906	Belfast Steamship Company	1869	18	804
Patriotic	1911	Belfast Steamship Company	2254	18	840
Vera	1898	London & South Western Railway	1100	18	510
City of Belfast	1893	Barrow Steam Navigation Company	1055	18	385
Duchess of Devonshire	1897	Barrow Steam Navigation Company	1265	18	345
Amsterdam	1894	Great Eastern Railway	1745	18	447
Berlin	1894	Great Eastern Railway	1745	18	447
Vienna	1894	Great Eastern Railway	1753	18	447
Dresden	1896	Great Eastern Railway	1805	18	476

Chapter 3 : Fast main-line passenger ferries – Harwich and South Coast

The abundance of fast triple and quadruple expansion engined steamers in service on the Irish Sea contrasts their paucity in the English Channel and in the North Sea. The North Sea is understandable as the longer crossings could not be traversed overnight even at speed. The Channel crossing is a bit more complicated and hinges on the change over from paddle steamer directly to turbine steamer in the Dover Strait without recourse to steam reciprocating engines. This left only the Newhaven to Dieppe and Southampton and Weymouth to the Channel Island services for the reciprocating engined steamer to excel.

*The **Empress** (1907) was one of a group of triple screw direct-drive turbine steamers that replaced paddle steamers in the Dover Strait.*

(Mike Walker collection)

At Harwich the Great Eastern Railway took delivery of their 17$\frac{1}{2}$ knot ferry **Chelmsford** in 1893. The three sisters of the **Berlin** class followed in 1894, each capable of 18 knots despite them having the same engine arrangement as the **Chelmsford**, and being slightly wider in the beam. Their bunker capacity was 150 tonnes of coal and they consumed this at a rate of 4 tonnes per hour at full speed. The **Berlin**, **Amsterdam** and **Vienna** came from Earle's at Hull and set new standards of luxury on the Harwich to Hook of Holland route. There was accommodation for 218 first class on the Main Deck and 218 third class aft in the poop and after 'tween decks. The boat train for the night service left Liverpool Street Station at 2000 hours to connect with the midnight departure from Harwich which arrived in Holland at 0600 hours.

The **Berlin** disaster is remembered to this day in Holland. Half an hour from her berth at the Hook in the early hours of 21 February 1907, the **Berlin** was swept in a north-westerly storm onto the north entrance pier of the New Waterway. The ship quickly broke up in the atrocious conditions and only 15 men survived.

The **Dresden** followed in 1896, also from Earle's, but she was lost to a submarine attack in 1918. The **Amsterdam** and the **Vienna** were eventually overtaken by the march of the turbine engine and retired in 1928 and 1930 respectively. The **Vienna** spent her last ten years as the **Roulers** on the Harwich to Zeebrugge route. The British partners on the service, the Great Eastern Railway, forsook the triple expansion engine at Harwich with the arrival of the turbine steamer **Copenhagen** and her sisters from 1907 onwards. The Dutch partners, the Zeeland Steamship Company, placed the **Oranje Nassau** in service in 1909, the **Prinses Juliana** in 1920 and the **Mecklenburg** in 1922, all with triple expansion engines and a service speed of 22 knots. The **Prinses Juliana** was lost in World War Two, but the **Oranje Nassau** survived on the route until 1954 and the **Mecklenburg** lasted until 1960, very much the last fast main line passenger steamer to operate in British waters.

*The **Vienna** lying at buoys off Parkeston Quay.*

(Author's collection)

From 1906 onwards, the Lancashire & Yorkshire Railway maintained the former Fleetwood steamer ***Duke of Clarence*** (see Chapter 2) on a summer-only service between Hull and Zeebrugge. The route was jointly maintained with the North Eastern Railway. In winter the ***Duke of Clarence*** (now registered at Goole) would either lay up or revert to secondary Irish Sea routes. However, she was rarely called upon to provide her full 19 knots and was generally scheduled for more economical speeds. Used as an Armed Boarding Vessel in the Great War, she was eventually replaced at Hull by her former partner from Fleetwood, the ***Duke of Connaught***. The ***Duke of Connaught*** served the route for three years between 1929 and 1932. The ships used to offer positioning cruises between Fleetwood and Hull usually via Scotland in spring and Jersey in autumn. These were always very well subscribed.

*The former Fleetwood steamer **Duke of Clarence** (1892) displaying ornate scroll work on a beautiful counter stern whilst alonside at Zeebrugge.*

(Author's collection)

The Newhaven to Dieppe route was operated by the London, Brighton & South Coast Railway. Although the company later developed very fast turbine steamers, the ***Seaford***, ***Sussex*** and ***Arundel***, all of which could make 20 knots, were of note. The ***Sussex*** was built as a replacement for the ***Seaford*** which was lost in her first year of service when she sank 20 miles off Newhaven inbound in fog, following collision with the French partner steamer ***Lyon***. The ***Arundel*** was delivered a few years later in 1900. She was very well appointed, with buffalo hide framed in teak and mahogany panelling adorning the first class smoking room on the Promenade Deck.

*The **Arundel** was the final 20-knot steamer to be built for the London, Brighton & South Coast Railway.*

(Author's collection)

*A rare photograph of the ill-fated **Seaford** (1894), probably on trials on the Clyde.*

(Mike Walker collection)

All three ships were equipped with twin four-cylinder engines. The cost of building the ships was £57 500 for the **Seaford** in 1894, £60 000 for the **Sussex** in 1896 and £68 000 for the **Arundel**. They were licensed to carry about 750 passengers in three classes. The **Sussex** transferred to the French flag in 1913. The **Sussex**, and in the 1920s the **Arundel**, was responsible for the twice-weekly summer excursion from Brighton Palace Pier to Dieppe and return via Newhaven and special train to Brighton. The **Arundel** was sold for demolition in 1934 having spent her last six years on relief duties for the younger turbine steamers.

On the Channel Islands routes, the steamers of the Great Western Railway, such as the **Roebuck** and **Reindeer**, and those of the London & South Western Railway, including the 1890-built **Stella**, **Frederica** and **Lydia** are fondly remembered, but none could achieve 20 knots in service. The **Roebuck** and **Reindeer** were built in 1897 at Barrow, and equipped with twin engines with 23, 36 and 56 inch cylinders and a 33 inch stroke. Their service speed was 19 knots and at trials they attained over 20 knots. They instigated a new summer only daylight service from Weymouth, spending much of the winter laid up at Milford Haven. One January night, the **Roebuck** caught fire at her winter moorings, and the weight of water needed to extinguish the blaze caused her to sink. She was refloated ten days later and was back in service in June that year.

*The **Roebuck** (1897), built for the Great Western Railway's Channel Islands routes.*

(Barrow Museum Service)

The winter service was maintained by a third fast steamer, the *Ibex*, which had been completed for the Great Western Railway by Laird Brothers at Birkenhead in 1891. She was involved in a number of incidents, the first on Good Friday 1897 whilst racing the *Frederica* into St Helier when the *Ibex* struck rocks off the Corbière Lighthouse. She managed to beach herself in Portelet Bay, around the corner, in a half submerged state. In January 1900, less than three years later, the *Ibex* grounded off St Peter Port in Guernsey and sank, to be salvaged six months later. She alone maintained the Channel Islands service from Weymouth during the Great War, and survived unscathed. The final incident was caused by a freak wave in July 1922 when panic spread as water gushed through the passenger accommodation, this time without injury. The *Ibex* was finally retired three years later.

The *Roebuck* was requisitioned for duties as an armed merchant cruiser during the Great War but was lost in Scapa Flow when she dragged her anchors across the bows of a battleship and sank in deep water. The *Reindeer* initially operated out of Fishguard with trooping duties, but late in 1914 was requisitioned as a minesweeper for use in the Mediterranean. Her first post-war duties were trooping between Weymouth and Cherbourg, and she returned to the Channel Islands in 1920. She was given a shelter deck aft of the bridge in 1923, along with an extensive refit to upgrade her facilities. Displaced by new turbine steamers, she was retired and broken up in 1928.

The *Frederica* and her two sisters were the first of the new generation of magnificent fast Victorian steamers built for the London & South Western Railway. It was these ships which laid down the challenge of the 1890s on the Channel Islands routes. They had a passenger certificate for 712 in two classes and berths for 42, and they maintained a service speed of 19 knots. A slightly slower pair was commissioned four years later for the Southampton to Le Havre service, the twin-funnelled *Alma* and *Columbia*, in service until sold in 1912. Their main distinction was the provision of individual cabins for the overnight services. A near-sister, the *Vera*, followed in 1898, although with only one funnel, for use on the Channel Islands, Le Havre and St Malo routes. She was hugely successful and her passenger configuration was altered several times during her career to suit the trade of the day. She survived until withdrawn in 1933.

Sadly, the one ship that is most often remembered is the last of the trio, the *Stella* due to her tragic loss in 1899. On a daylight sailing to Jersey on 30 March, the *Stella* hit the Casquets at full speed, the order having just been given to slow down due to fog. Only 112 out of the estimated 217 passengers and crew aboard the ship survived (the passenger list went down with the ship). The incident did, however, end competitive racing on the route and led to the joint service with the Great Western Railway. Always a problem in the Channel Isles, fog is still a hazard and even now is the reason for more cancelled and postponed flights to and from the islands than any other cause.

The *Stella* had left Southampton on Maundy Thursday at 1120 hours after the arrival of the Waterloo boat train. She was under the command of Captain William Reeks and the mate was Richard Wade. Her course was set by dead reckoning on leaving the Needles for a point west of the Casquets, which are situated eight miles west of Alderney, en route for St Peter Port. Normal practice was to judge the distance run from engine revolutions rather than from a patent log. The ebb tide was running on her port beam at an angle just south of west, however, the tide may have been slacker than expected. An additional risk was that the ship had just been overhauled and was running with a clean bottom, and finally there was a doubt as to whether the compass deviation cards had been rechecked after the overhaul. Her three compasses were known to have been last adjusted two years previously.

Running into fog two hours short of Guernsey, the ship unknowingly was placed at risk. The longer that Captain Reeks dared to maintain full speed in competition with the Great Western Railway steamer also running for St Peter Port, the greater that risk became. Whatever the cause or combination of causes, the ship hit the Black Rock at full speed just as the order to reduce speed had been issued. Why the ship was east of her course and why she had arrived there earlier than calculated will never be known, but the cards were stacked against her.

At the Board of Trade Inquiry it was estimated that the *Stella* sank in about eight minutes. Five boats were got away and the sixth overturned. There was no reported confusion or panic with the crew helping passengers to get away first. As the ship's boilers became submerged there was 'an awful report and loud thuds, then hissing sounds and a blinding hiss of steam'. Heroism was the rule of the day: stewardess Mary Rogers gave her lifebelt to a lady passenger, refused to board one of the boats for fear of swamping it and went down with ship. As the *Stella* settled lower in the water, the Reverend George Clutterbuck led a group in prayer and as the bow reared upwards and the stern disappeared, the Captain remained steadfast on his bridge.

The Inquiry concluded:

The *Stella* had not made good the course set and her master had continued at full speed in thick weather when he must have known his vessel was in the immediate vicinity of the Casquets, without taking any steps to verify his position.

And as for racing to port the Inquiry could not acknowledge this as a cause of the accident and concluded:

When different lines compete to the same ports, rivalry will naturally exist.

Of her sisters, the **Frederica** was displaced and sold to Turkish owners in 1911, and was lost in operation as a minelayer in 1914. The **Lydia** had to be beached at the head of St Peter Port Harbour after hitting La Rond Rock at the age of one. Ten years later she was on fire at St Helier following a down draught to her furnace. In the war she was narrowly missed by a torpedo, but was returned to civilian duties in 1919 after trooping between Southampton and St Malo. Sold as surplus to requirements, the **Lydia** led a nomadic life, including a period of inactivity at Ardrossan under Coast Lines Limited management, until demolished in 1933, following service under the Greek flag.

The last of the triple expansion steamers was the **Alberta**, completed in 1900 by John Brown & Company. She was ordered as a replacement for the **Stella**. Capable of outpacing the turbine steamers that displaced her in 1930 she was sold for use in Greek waters only to be sunk in 1941 during an air raid. Like the **Vera**, she underwent a variety of modifications during her career in order to keep her up to date. Of all the Channel Island steamers, the **Alberta** was the most imposing, with straight stem, raised bridge and counter stern, a tall single funnel and a long Promenade Deck.

Perhaps the most striking aspect of the crack Victorian and Edwardian steamships on the cross-channel routes was the number of accidents that befell them. Acts of war apart, the number of vessels that foundered or were lost at sea remained at a level that was not acceptable, but which the succeeding era of the fast turbine steamers was about to rectify. That apart, these ships were the pinnacle of the steam reciprocating engine's career at sea, crewed by the nation's finest and revered by the travelling public. The latter, it should be added, were the keenest followers of any race that might ensue between rival companies!

*The accident-prone **Lydia** (1890), of the London & South Western Railway fleet, spent some time laid up at Ardrossan under Coast Lines management.*

(Author's collection)

Chapter 4 : The intermediate ferries

There were numerous steam reciprocating engined intermediate ferries in the Edwardian period. Their duty was to run from port to port, generally overnight, and at a modest speed, to deliver passengers and goods to their destination for the next day. They were common on the longer routes of the North Sea and Irish Sea, but virtually absent on the English Channel where only fast crack ships would suffice.

Probably the least known, but nevertheless particularly interesting group of intermediate type ferries, were the emigrant ferries of the Wilson Line. The Wilson Line had carved a very lucrative niche in the early twentieth century ferrying emigrants from Sweden and Norway to Hull thence by special train to Liverpool to connect with British liner services for onward voyages to the New World. A deal had earlier been struck with the Hamburg America Line not to run cheap services from Germany in return for the German liner company not extending their routes to include Gothenburg. That being so, the emigrant trade diminished in the early 1920s with the US Quota Act in 1921 and the Johnson Reed Act which followed in 1924.

John Harrower, in his book *Wilson Line* (see reference section), described the early days of the emigrant ships:

The carriage of emigrants continued until the outbreak of war in 1914, all Wilson passenger ships being fitted with accommodation for them in the 'tween decks. The ***Romeo*** built in 1881 is credited with being able to carry one thousand when employed on the Gothenburg service. The ***Eldorado*** of 1886, which ran from Hull to Stavanger and Bergen, had accommodation for some 250 Norwegian emigrants, the first Wilson ship to be fitted with electric lighting, she doubtless improved conditions in the 'tween decks for hitherto oil lamps, in no great number for fear of fire, had provided only limited illumination.

Major improvements to the accommodation were made in the ***Calypso*** which was delivered from Earle's Shipyard and Engineering Company at Hull in 1904. She had a series of eight berth cabins for steerage passengers, improved toilet facilities and even a small smoke room. However, there was still 570 open berths in the forward 'tween decks on offer as a cheaper option. The ***Calypso*** also had superior berths for 45 first class and 46 second class passengers. A paltry 2876 tons gross and with triple expansion engines giving only 283 nominal horse power, the 'tween decks were an uncomfortable and no doubt unpleasant place on a rough winter's night. The ship was torpedoed and sunk with the loss of 30 lives in 1915.

Earle's built a variety of vessels fitted out for the emigrant trade, each to a different specification and design. The ***Oslo*** was delivered in 1906 with accommodation 'for about 400 other steerage in temporary berths'. She too was lost to a submarine in the Great War. The ***Aaro*** was delivered in 1909 with accommodation for 104 first class, 40 second class and 50 steerage class, and was also lost to a German submarine. She was followed in 1910 by the ***Eskimo***, distinctive in having twin funnels and quadruple expansion engines – captured in the war and never refitted by her owners on her return, and finally the ***Bayardo*** which was delivered in 1911. The latter broke her back on the Middle Sands in the Humber after running aground in dense fog in January 1912 – at the tender age of 6 months. All on board were taken off safely.

*The six-month old **Bayardo** (1911) lies across the Middle Sands in the Humber with a broken back in January 1912. She was on passage from Gothenburg. All passengers and crew were safely taken off and much of the cargo and fittings salvaged before the wreck was blown up.*

(Author's collection)

Replacement tonnage was the second-hand *Orlando* built in 1904 for Rennie's Aberdeen Line, and the *Rollo* built in 1899 for the African Steamship Company Limited. Both were sold for demolition in 1932. In 1922 Ramage & Ferguson of Leith delivered the new triple expansion engined *Tasso* to the Wilson Line. She had comfortable accommodation for 45 first class and 362 third class passengers, the term steerage class being dropped after the war. Her arrival in the fleet coincided with the reduction in emigration to the United States and she was something of a white elephant. She was sold seven years later, but survived until 1941 when she was scuttled as a block ship at Massowah. Thereafter, many of the Wilson Line North Sea ships continued to offer limited first and second class accommodation, but the passenger service became very much subordinate to the cargo that was on offer.

Andrew Weir & Company also recognised the value of the Baltic and Scandinavian transmigrant trade through Hull, as also did the Finland Steamship Company. In 1919, and jointly with the Danish East Asiatic Company, Andrew Weir set up the Baltic Shipping Company, later to become the United Baltic Corporation. The first three ships were chartered from the East Asiatic Company: *Baltannic*, *Baltabor* and *Baltriga*. Each could carry 130 passengers in three classes. Elder Dempster Lines' *Akabo* was bought in 1926 to provide greater passenger accommodation. As the *Baltonia*, she could carry 80 cabin class and 130 third class passengers. This ship was responsible for bringing hundreds of members of the Polish Jewish community to safety in Britain before the outbreak of World War II.

A number of ships were bought into the company in the 1930s including the former General Steam Navigation Company's sisters *Heron* and *Starling* which became the *Balteako* and *Baltallin*, each with comfortable accommodation provided in three two berth cabins. These coal burners had a speed of 12 knots and a modest tonnage of just over 1300 gross tons. They were built in 1920. Another big passenger carrier joined the fleet in 1935 in the form of the Furness Withy Bermuda & West Indies Steamship Company liner *Dominica*, which had originally been built for their Canadian services as the *Digby* in 1913. As the *Baltrover*, she could carry 194 first and third class passengers and remained on the Baltic services until 1939. Most subsequent purchases were diesel vessels and few had capacity for more than a handful of passengers. The post-war 14 knot, single screw steamer *Baltrover*, however, built in 1949 as *Marstenen*, and which joined the otherwise largely diesel fleet of United Baltic Corporation in 1950, carried only four passengers on her usual passage between Hayes Wharf and Gdynia. She was sold in 1968 and became a fire casualty whilst on passage from Dubai to Karachi in 1976.

Better known intermediate class ferries are ships operated by the railways, by companies such as the North of Scotland, Orkney and Shetland Shipping Company and J & G Burns, later Burns & Laird Lines Limited.

The Great Eastern Railway maintained a regular passenger service between Harwich and Antwerp until the advent of the Great War. The steamers were similar to those on the faster Harwich to Hook of Holland service but were designed with more modest service speeds. The last ship built for the route was the *Brussels*, which was delivered by Gourlay Brothers at Dundee in 1902. The Great Eastern Railway traditionally bought ships from Earle's Company at Hull, but this shipyard had just been saved from bankruptcy by Thomas Wilson and had a full order book for the Wilson Line. The *Brussels* featured 160 first class berths and a special suite of rooms for the ladies. Although sunk as a block ship at Zeebrugge in the Great War, she was later raised and ended her days carrying cattle between Dublin and Preston during the 1920s.

*The **Brussels** shows a clear similarity to the faster Hook of Holland steamers operated by the Great Eastern Railway.*

(Mike Walker collection)

More typical of the railway ships was the quintet of steamers delivered by Earle's Company at Hull for the Great Central Railway in 1910 and 1911 at a cost of approximately £41 000 each. They were described by Duckworth and Langmuir (see reference section) as typical passenger and cargo ships of about 1600 tons with triple expansion engines. The first three, the *Dewsbury*, *Accrington* and *Blackburn* were delivered in 1910 for the Grimsby to Hamburg Royal Mail service, with a fourth, the *Bury*, due for delivery the following year. The loss of the *Blackburn* off the Norfolk coast in December 1910, happily without loss of life, required a replacement vessel, the *Stockport*, to be laid down in January for delivery in the summer.

The ships had comfortable accommodation for 140 first class passengers amidships, 10 second class in the poop and 300 steerage passengers in the forward 'tween decks. Public rooms included a saloon which occupied the full width of each ship on the Main Deck, a gentleman's smoking room and a lady's lounge on the Bridge Deck. They had space for 4000 cubic metres of cargo and could maintain a speed of just over 14 knots.

The *Dewsbury*, *Accrington*, *Bury* and *Stockport* were distinguished in two ways. First they all served as convoy rescue ships in the Second World War, the *Stockport* being lost in this capacity in 1943 (see Chapter 5). The second was their longevity. They became part of the mighty London and North Eastern Railway in 1923 and then passed into Associated Humber Lines in 1935. The company brochure for 1936 recounts:

Steaming across the romantic North Sea, mile after pleasant mile, is the ideal approach to Germany . . . Late some afternoon you arrive at Grimsby 'to take the boat to Hamburg'. What joy what excitement there is behind those formal words . . .

And we thought the spin doctor was a recent invention! They were comfortable ships with wood panelled saloons, silver plate engraved with ship and company names, and dining saloon chairs engraved with the letters GCR on their backs. The ships were upgraded in the late 1930s after which the company brochure included the description:

Wash hand basins with hot and cold running water have been installed, and electric light is fitted throughout. Special facilities have been provided for handling motor cars promptly and safely.

The *Bury* and *Dewsbury* served until 1958 and 1959 respectively, with the *Accrington* withdrawn in 1951. In post war years the Grimsby station was closed and the *Bury* partnered the Associated Humber Lines' *Melrose Abbey* on the Hull to Rotterdam service while the *Dewsbury* was transferred to Harwich.

The *Melrose Abbey* was a charming ship, again built by Earle's Co at Hull and delivered in 1929 to the Hull & Netherlands Steamship Company. A coal burner with a single screw she could attain 14 knots, burning one tonne of coal per hour at 13 knots and $1\frac{1}{2}$ tonnes at full speed. She could accommodate 84 first class passengers in the central part of the ship, and as was common in those days, the 38 steerage class occupying the poop, the latter, of course, farthest from the life boats! The *Melrose Abbey* also saw distinguished service as a convoy rescue ship during the war along with the *Bury* and her sisters.

The post-war service, with four sailings a week to Rotterdam, transferred to the Humber Dock until 1959 when the Riverside Quay was reopened to traffic after being damaged in the war. In June 1958 the *Bury* was replaced by the new diesel ferry *Bolton Abbey*, and the old *Melrose Abbey* ran alongside the new ship until a new *Melrose Abbey* displaced her at the end of the year. Thus the last steamers of the former Great Central Railway and the Hull and Netherlands Steamship Company left the Humber. The steamer *Melrose Abbey*, however, went on to enjoy a second career as a cruise ship in the Mediterranean, finally converted to burn oil fuel but retaining her triple expansion engines, and refitted to carry 282 cruise passengers.

Whilst others invested heavily in turbine steamers, companies such as Burns & Laird Lines Limited and its predecessors remained faithful to the reciprocating engine. The early history of the combined fleet always appears complicated but a synopsis to explain some of the familiar names involved is useful. At the amalgamation of J & G Burns Limited and the Laird Line in 1922, Burns contributed the passenger and cargo ships *Hound*, *Pointer*, *Magpie*, *Vulture*, *Woodcock*, *Partridge*, *Moorfowl*, *Puma* and *Tiger*, and Laird the *Thistle*, *Olive*, *Lily*, *Rose* and *Maple*. Laird's *Brier* and *Dunure* had limited passenger accommodation.

*The **Melrose Abbey** (1929) was displaced from the Hull to Rotterdam service in 1958.*
(Bernard McCall collection)

The ships of the two companies had quite distinctive characteristics. The Burns ships had fine lines, well raked masts and funnels with a single derrick on the after side of the foremast and another forward of the mainmast; the saloon accommodation was aft and steerage was well forward. The Laird ships were more stately in appearance with shorter derricks fore and aft of the foremast and a short derrick aft of the mainmast. They were altogether broader vessels and accommodation was provided amidships for saloon passengers and aft for steerage.

Nearly all the ships were powered by triple expansion engines driving a single screw and most could attain 15 knots. The *Brier* had twin cylinder compound engines and an iron hull. G & J Burns also had the direct-drive turbine steamer *Viper* on the daylight Ardrossan to Belfast service until 1920 when she was sold to the Isle of Man Steam Packet Company.

The combined fleet continued to serve the traditional Irish Sea passenger routes between Glasgow and Greenock and Londonderry, Glasgow and Belfast and Glasgow and Dublin, and Heysham to Dublin and Londonderry, the Heysham passenger services being discontinued in the early 1930s. However, the amalgamation allowed some of the older units of the fleet to be withdrawn and in 1929 the remainder of the fleet was given new names as follows:

Pointer, built 1896, became *Lairdsvale*
Magpie, built 1898, became *Lairdsgrove*
Vulture, built 1898, became *Lairdsrock*; sold in 1937 to David MacBrayne Limited and
    renamed *Lochgarry*
Woodcock, built 1906, became *Lairdswood*
Partridge, built 1906, became *Lairdsloch*
Moorfowl, built 1919, became *Lairdsmoor*
Puma, built 1899 as *Duke of Rothesay* for the Dublin & Glasgow Sailing and Steam Packet
    Company; later became *Lairdsford*
Tiger, built 1906 as *Duke of Montrose* for the Dublin & Glasgow Sailing and Steam Packet
    Company; later became *Lairdsforest*
Olive, built 1893, became *Lairdsbank*
Lily, built 1896, became *Lairdspool*
Rose, built 1902, became *Lairdsrose*
Maple, built 1914, became *Lairdsglen*
Brier, built 1882, became *Lairdsoak*

*The Laird Line's **Olive** was later renamed **Lairdsbank**.*

(P&O Group)

The cargo only vessels were similarly treated. The older members of the fleet were weeded out during the 1930s and at the outbreak of war in 1939 the fleet comprised **Lairdsgrove**, **Lairdsrose** and **Lairdsglen**, on the Derry run, with additions brought in during 1930 from the British & Irish Steam Packet Company: **Lairdshill**, formerly **Lady Longford** and built as **Ardmore** (sister to **Kenmare**, see below), **Lairdscastle**, formerly **Lady Limerick**, and **Lairdsburn**, formerly **Lady Louth**, and dating from 1921, 1924, and 1923 respectively. The **Lady Louth** and **Lady Limerick** had berths for 80 first and 90 steerage passengers, whilst the **Lady Longford** had fewer berths. As the three Ladies on the overnight Liverpool to Dublin route, they had carried notices on the Promenade and Boat decks which warned:

Passengers are reminded that noise created by walking and talking on deck after 11 pm is disturbing to those that have already retired to their cabins.

The **Lairdscastle** and **Lairdsburn** took over the overnight Belfast service until they were displaced to the Dublin route by new diesel tonnage in 1936. These two steamers had vastly superior accommodation to the remainder of the fleet in the early 1930s. The **Lairdscastle** was lost early in the war but her sister served the Dublin route after the war until withdrawn and scrapped in 1953. The **Lairdshill** served the Dublin route in the 1930s and Derry after the war alongside the motorship **Lairds Loch**. The **Lairdshill** was withdrawn in 1957. The surviving pre-war Derry ships were scrapped soon after the war.

*The **Lairdsgrove**, alongside at Glasgow, was formerly the Burns' steamer **Magpie**.*

(P&O Group)

*The Dublin & Glasgow Sailing and Steam Packet Company's **Duke of Rothesay** (1899) ended her days as **Lairdsford**.*

(Author's collection)

But what were these ships like to travel on? The thought that the old *Rose* could carry 140 first class and 700 steerage passengers and the *Olive* 1000 third class passengers on the long and exposed passage to Derry, evokes images of crowded and unpleasant conditions in peak season when cattle were replaced by steerage passengers in the 'tween deck accommodation! The older ships had the traditional passenger layout with first class accommodation aft, and with a saloon with swivel chairs around a single large table and some settees. Cabins offered four berths and ladies and gentlemen also had small dormitory type sleeping accommodation.

Lawrence Liddle in his book *Passenger Ships of the Irish Sea* (see reference section) describes the early accommodation:

Sometimes these ships had so called smoking rooms, small apartments either aft, incorporated in the saloon entrance, or on the midships Boat Deck, at the after end of the engine room skylight. There was rarely any other public rooms though, if the smoking room was on the Boat Deck, the space ahead of the companionway to the saloon might be designated as the lounge. Berths could be, and at busy times were, made up in the public rooms. Third class passengers were very roughly catered for in the 'tween decks forward, in close proximity to the livestock pens. These old ships had no below deck ventilation, other than what air came through the coal ventilators. The smell, which was compounded by cattle, coal smoke, vomit, metal polish, hot oil and cooking is not easily forgotten.

*The **Lairdsrose** (1902), formerly the **Rose**, carried up to 1000 passengers on the Glasgow to Derry route.*
(P&O Group)

The later steamers, such as the *Woodcock* and *Partridge*, built from the Edwardian period onwards had gradually improving standards of accommodation. Generally the dining saloon was at the fore end of the superstructure with the first class cabins aft. There was often a lounge in a deckhouse below the bridge along with the first class entrance; this area was enlarged by plating in the adjacent deck space as was the case on the *Lairdsglen*. Some ships had a small smoking room and bar at the after end of the Boat Deck. Third class passengers enjoyed the run of the poop, but multi-berth cabins were still not generally available. There was, however, usually a small, steamy and crowded bar. Cattle remained an important part of the economics of the steamers, and the animals and their drovers occupied the 'tween decks with dockside access via large hinged doors through which the animals embarked and disembarked encouraged by sticks with metal tips.

Perhaps of all the Burns and Laird ships the *Magpie/Lairdsgrove* is the most celebrated. She was the first Burns steamer to be employed on the Dublin service. Transferring from Derry to Dublin in 1908, the *Magpie* operated an express service from Glasgow every Tuesday at 2000 hours and returned at the same time on Thursdays from Dublin. The voyage took 18 hours and a first class berth in a two or four berth cabin cost a supplement of 2/6d. At the age of 25 she was reboilered and converted to burn oil fuel, and at the same time her accommodation for both passengers and crew was rebuilt and upgraded. She maintained a cargo only service between Glasgow and Derry in the two wars and in 1948 at the grand age of 50 was withdrawn and scrapped.

On the Derry run, the *Rose/Lairdsrose* is to this day remembered as the finest and steadiest ship, and the *Maple/Lairdsglen* is remembered as having an unpleasant tendency to roll. The old *Magpie* is barely remembered at all, but this was a ship which had earned her keep successfully for half a century and had repaid her owners time and time again.

One other important group of intermediate ferries in terms of innovation and design were the City of Dublin Steam Packet Company's county class. They maintained the company's twice daily Liverpool to Dublin service with an eight hour crossing at a speed of 14 to 15 knots. The *Cork* was lost in the Great War and the remainder transferred to the British and Irish Steam Packet Company in 1919 with the suffix Lady added: *Lady Louth*, *Lady Wicklow*, *Lady Carlow* and *Lady Kerry*. The *Kilkenny* was sold to the Great Eastern Railway and renamed the *Frinton*. The *Louth* survived until 1938 and the *Wicklow* until 1949.

*The Liverpool to Cork service was maintained by the coal-burning **Kenmare** until 1956.*

(Author's collection)

In the southern Irish Sea, the City of Cork Steam Packet Company maintained an overnight service between Liverpool and Cork, usually via Fishguard. From 1921 onwards the service was maintained by the *Kenmare* and the *Ardmore*, although the latter was quickly transferred to the British and Irish service between Dublin and Liverpool as the *Lady Longford* (see above) and then to Burns and Laird as the *Lairdshill*, running between Dublin and Glasgow until withdrawn in 1957. The *Kenmare* stayed on the Cork route throughout her career but used Fishguard only as her eastern terminal throughout the war years. In the 1930s she would leave Liverpool on Thursday morning and arrive at Cork on Friday morning, departing again on Saturday in the late afternoon, arriving at Liverpool after calling at the cattle lairage at Birkenhead some 24 hours later. There was sufficient lay over at Liverpool to allow an extra cattle run if required. Passengers were not carried in the war years, but the *Kenmare* reopened the passenger service between Cork and Fishguard in August 1945, the previous incumbents, the *Innisfallen* and *Inniscarra* being war losses; the *Kenmare* only returning to the Liverpool service in 1948. The *Kenmare* was coal-fired whereas the *Ardmore* was oil-fired.

The *Kenmare* could accommodate 60 saloon passengers in the central superstructure and additional steerage passengers in the poop. The main dining saloon was directly beneath the bridge on the Lower Deck and there was a general room above on the Promenade Deck which offered comfortable settees for unberthed passengers, many preferring the lounge to the cabins which tended to be hot and stuffy and some cabins suffered also from engine noises. Her last return sailing from Liverpool to Cork took place in May 1956 when she was displaced on the route by the 60-passenger *Glengariff*.

The very last of the intermediate ferries to be built for the Irish Sea was the *Great Western*. Built for the Great Western Railway's Fishguard to Waterford service by Cammell Laird in 1934, the new ship was a replacement for the *Great Western* of 1902 which at one time partnered the route with sister *Great Southern*. The new ship was powered by twin fast running triple expansion engines fed by automatically fired coal burning water tube boilers. She maintained a service speed of 16 knots, much as her predecessors had done. Although principally a cargo and cattle carrier with stalls for 668 cattle, she also had comfortable accommodation for 450 passengers.

During the war she acted for a time as a troopship and distinguished herself by driving German bombers away with her own guns whilst off the Wexford coast. Returning to her peace time duties, she eventually clocked up 5100 voyages on the overnight Fishguard to Waterford route and was withdrawn and sold for scrap in 1967 after a brief period on the Heysham to Belfast cargo route. Earlier in 1959 British Railways had withdrawn the catering staff and the ship had become cargo only. Now converted to burn oil fuel her service speed was reduced to just 14 knots. She was nevertheless the last of the intermediate type ships to trade on the west coast, and with the sale of the former *St Clair* (see Chapter 7) also to the breakers in 1967, it was truly the end of the slow overnight steam passenger ferry.

*The **Great Western** (1934) on cargo duty at Heysham Harbour in April 1967 at the end of her career. She was sold for demolition in Belgium later that year.*

Chapter 5 : Coastal liners

A popular alternative to rail travel up and until the Great War was the coastal liner. Cheaper than the train, a coastal voyage offered the chance to relax for a few days on the journey to and from London from places such as Dublin, Liverpool, Newcastle or Leith. After World War One, some services were revived but the coastal liner was now in competition with cheaper travel by road and the motor coach, and the cruise element of the first class clientele was finally eroded by the onset of the Great Depression. Although some services continued through the 1920s and 1930s, none survived after the Second World War save for the twelve passenger diesel ships of Coast Lines Limited on the Liverpool to London route.

Generally capable of between 13 and 15 knots, the coastal liner offered comfortable accommodation for first class passengers with slightly more austere second class facilities. Trading between their home port and London, there might be intermediate calls such as Plymouth out and Falmouth return plus Southampton and Dublin on the run between Liverpool and London. For many years this was the duty of the Coast Lines' steamer ***Southern Coast***, built originally for Hough Line as the ***Dorothy Hough*** in 1911. She became ***Southern Coast*** on the formation of Powell, Bacon and Hough Lines in 1913 (which traded as The Coast Line, becoming Coast Lines Limited in 1917). She could carry 80 passengers and sailed on a two-week round trip, loading and unloading cargo as required. Displaced by a pair of 12-passenger diesel units in the mid 1930s (the ***British Coast*** and ***Atlantic Coast*** which were commissioned for the service in 1933 and 1934 respectively), the ***Southern Coast*** was sold to the Falkland Islands Company, later to be lost to a war time mine in UK waters.

*The **Southern Coast** (1911) from a series of art cards issued by Coast Lines during the 1930s.*

(Author's collection)

Other passenger carrying steamers on the west coast were operated by the British & Irish Steam Packet Company. Founded in 1836 to maintain a reliable service between Dublin and London, the route was at its most popular just before the Great War when there were four sailings per week in each direction. The last of the iron-hulled compound engined ships was the ***Lady Olive*** dating from 1878, and the first steel-hulled triple expansion engined vessel was the ***Lady Martin*** completed in 1888. At the outbreak of war the newest steamers were the ***Lady Roberts*** built by Ailsa at Troon in 1897 and the ***Lady Gwendolen*** built by the Clyde Shipbuilding & Engineering Company at Port Glasgow in 1911. They could complete the voyage to London with intermediate stops at a variety of ports in four days at a speed of 13 knots. They had accommodation for 120 saloon and 50 second class passengers. The route was advertised under the banner Grand Summer Cruises, and David McNeill in his book Irish Passenger Steamship Services explains:

Single fares were 27/6d saloon and 19/6d second class. Food was not included but passengers could either pay for each individual meal or buy a contract ticket for all meals at 23/-d saloon or 17/-d second class. Guinness stout was available at 4d a bottle and a basin of soup cost 4d, the latter beverage only available to second class passengers.

The last steamers to be built for the Dublin to London service were the ***Lady Wimborne*** which came from the Clyde Shipbuilding and Engineering company and the ***Lady Cloé*** which was built by Sir Raylton Dixon at Middlesbrough, both in 1915. They had accommodation for 70 first class passengers only, and at the end of the war resumed the passenger service with intermediate calls at Cork, Falmouth Torquay and Southampton against a £5 single passage ticket. The three other ships on the route then no longer offered passenger accommodation. Popularity was never regained, and in the 1930s there was only one sailing per week, and the passenger accommodation was withdrawn altogether in 1933. After the Second World War the route was incorporated into the Coast Lines' Liverpool to London service operated by the 12 passenger diesel vessels. Of all the Lady ships, the **Lady Wimborne** fared best, only being sold for demolition in 1955, having become *Galway Coast* in 1939 and sold out of the Coast Lines group in 1945

William Sloan & Company carried passengers on their Glasgow-Belfast-Bristol route until 1932, when it became cargo only. The last of the passenger steamers were the **Findhorn** built in 1903, **Annan**, built in 1907, and the sisters **Beauly** and **Brora** of 1924, all of which came from Ailsa at Troon. All four ships survived well into the 1950s.

*William Sloan's **Annan** (1907) seen in the Avon approaching Bristol.*

(Author's collection)

The Clyde Shipping Company entered the coastal passenger trade in 1856 and by the 1900s had a daily departure from Springfield Quay at Glasgow for London, Southampton, Belfast, Dublin or Waterford. In the 1930s the round trip fare was 60/-d, meals extra, for a Glasgow to Dublin, Waterford and Cork round trip lasting five days. The last steamers to be built for the company, and the last ever steam driven coastal liners were the **Beachy** and **Rathlin**. The former was lost in the Second World War whereas the **Rathlin** went on to become the **Glengariff** on the City of Cork Steam Packet Company's Cork to Liverpool service, following three years working under the Burns & Laird banner as the **Lairdscraig**, when she offered saloon accommodation for 40 passengers. The **Glengariff** took over from the venerable **Kenmare** in 1956. The **Glengariff** closed the 140 year old Cork to Liverpool passenger link in 1963 and was then sold for demolition. The ships had open bridges until World War Two, and it is said that the Deck Officers were easily recognised ashore by their weather-beaten appearance!

Frank Bowen described the two ships in an article in the *East Ham Echo* and later incorporated in his book *London Ship Types* (see reference section):

The two steamers **Rathlin** and **Beachy**, which were added to the Clyde Shipping Company's fleet in 1936, may be quoted as typical of the best of the type. They were 270 feet (82 metres) long by 38 feet (12 metres) broad, are magnificent sea boats, and have comfortable accommodation for 42 first class passengers in cabins with a number of second class passengers in the 'tween decks forward. Triple expansion engines give them a speed of $13^1/_2$ knots.

An earlier pair was the 13 knot **Toward** and **Copeland** built in 1923 with accommodation for 40 first and 27 second class passengers. The **Toward** was lost in the war as a convoy rescue ship but her sister became the **North Down** for North Continental Shipping Company Limited in 1947, and was used on her owners passenger and cargo service from Belfast to the Continent although soon relegated to just a cattle carrier. She ended her days as the **Ulster Herdsman**, having briefly been owned by the Blue Star Line (Vesteys) as **Drover**. As the **Ulster Herdsman** she operated on general duties for the Belfast Steamship Company until withdrawn and scrapped in 1963. She remained a coal burner to the end.

*Literally the end of the line for the **Ulster Herdsman** (1923), formerly Clyde Shipping Company's **Copeland**, seen at the end of her career in May 1963.*

Anther of the company's ships which was used in the war as a convoy rescue ship was the ***Goodwin***, one of a pair built in 1917, her sister, the ***Longships***, being lost by stranding off Scilly in December 1939. In the early 1930s they were popular with passengers in summer on the West Coast run, and Captain Blair would rig up one of the small deck cranes as swings if small children were aboard. Interestingly radio and radio officer were only carried in the summer months when passengers were aboard. Radio Officer Roberts reported in a letter to the Editor of *Sea Breezes* in January 1981:

Even in 1936 radio was a bit of a mystery to some, and I can recall one lady handing me a message to send. It was tightly folded up and when I started to unfold it to check the words, she snatched it out of my hands and called me a very inquisitive young man, and she would report me to the captain (which she did). I never learned what was in the message but believe that she thought the ship's aerial was some kind of catapult!

*The **Goodwin** (1917) was one of a number of coastal liners to be used as Convoy Rescue Ships in the war.*

(National Maritime Museum)

The 1939 Clyde Shipping Company timetable offered the following inclusive first class round trip fares from Glasgow: 7 day trip to Belfast, Waterford and Plymouth £4 5/-, 9 day trip to Belfast and London £5 5/-, 6 day trip to Dublin Waterford and Cork for £3 5/-. Similar fares were offered for a variety of round trips from London changing ship for the return voyage. An attractive round trip was offered for £4 10/-, north to Glasgow by Clyde Shipping Company steamer on Tuesday from London. The return was from Dundee on Wednesdays by Dundee, Perth and London Shipping Company vessel arriving London on Friday some 34 hours later or from Leith by London & Edinburgh Shipping Company steamer on Monday, Wednesday or Saturday. The same timetable advises that Breakfast is served at 8.30 am, Dinner at 1 pm and Tea at 6.30 pm. A bottle of beer was 9d and a whisky was 1/8d; champagne, it appears was sold by the pint bottle at 10/6d a time!

On the East Coast North Sea services, the fastest ships were the Carron Company steamers *Avon* dating from 1887 and *Carron* built in 1909. They were big ships, the *Carron* having a gross tonnage of 2351, and could complete the express voyage between London and Grangemouth in 28 hours at a speed of 15 knots. They could carry 600 passengers, 100 of them in first class, the remainder in steerage. The passenger service ceased after the Great War.

The last steamers on the London and Edinburgh Shipping Company's passenger service were the Caledon-built *Royal Scot* which dated from 1910, and the *Royal Fusilier* and the *Royal Archer*, which had been built in 1924 and 1928 respectively. A new ship of the same name replaced the *Royal Scot* in 1930, but this one had first class accommodation only for twelve passengers in two berth cabins. The old *Royal Scot* became the relief vessel *Royal Highlander* but was sold in 1932. The *Royal Fusilier* and the *Royal Archer* had capacity for 160 first class in two and four berth cabins amidships, and a further 32 first class passengers could be accommodated in temporary berths in the dining saloon, smoke room and music room on the Bridge, Main and Promenade decks respectively. Berths for 130 second class passengers were aft, in four, six or eight berth cabins, with 14 extra in the dining saloon if necessary and two ten berth dormitories. Passengers had access to the broad Promenade Decks, from where they could enjoy the afternoon sail from Hermitage Steam Wharf down the Thames, with two nights at sea before a dawn arrival at Leith on the third day. These were the only ships in the fleet to have Royal names.

*The **Royal Archer** (1928) was the last passenger ship to be built for the London to Leith service.*

(National Maritime Museum)

The *Royal Fusilier* and *Royal Archer* were slightly slower than Carron's *Avon* and *Carron* and had a service speed of just under 14 knots, but were great favourites on the East Coast. The regular service was interrupted in 1937 when the *Royal Fusilier* sank off Gravesend after a collision, but it resumed shortly afterwards after the ship was raised and refurbished. The service ceased in 1939; sadly none of the company's ships survived the war beyond 1941.

The last steamer on the Dundee route of the Dundee, Perth and London Shipping Company was the *Perth*. She was built by Caledon in 1915 and she too was big with a gross tonnage of 2248. Capable of 15 knots, the *Perth* had comfortable accommodation for 60 first class and 72 steerage class passengers, although she initially saw service as an armed patrol boat. She maintained the East Coast passenger service to London until 1939, latterly alongside the 12 passenger *London*, but the service was not resumed after the war. Berthing at Limehouse in London, the *Perth* called also at Southend Pier from 1932 onwards to collect and disembark passengers to and from the 30 hour run up to Dundee. This could save passengers up to four hours on their journey to and from central London, but no other coastal liners used this facility. She typically left the Thames every Wednesday and Dundee every Saturday. The ship was sold after the War and saw service initially with the Falkland Islands Company and later under the Italian flag. She was scrapped in 1962.

Aberdeen also had a direct connection to London maintained by the Aberdeen Steam Navigation Company. This service was operated by the *Aberdonian* which had been built in 1909 by D & W Henderson & Company at Glasgow and the *Lochnagar*, built in 1906 as the *Woodcock* for G & J Burns (see Chapter 4). The *Aberdonian* carried 80 first class and up to 200 steerage class passengers at a stately 13 knots. Whilst coming up the Thames to Limehouse, those passengers needing to get quickly to the city were given the option of disembarking at Gravesend by launch.

The company was bought by the Tyne Tees Steam Shipping Company in 1946, Tyne Tees having been acquired by the Coast Lines group in 1943. The *Aberdonian* survived two world wars, for much of the time as a coastal forces depot ship. Her master was presented with a silver cigarette box by the mayor of Aberdeen when the ship was requisitioned for a second time at the start of World War Two.

The Aberdeen to London service did resume briefly in 1947 with a new build - the 12-passenger diesel-driven *Aberdonian Coast*. The service was not successful and the ship became the *Hibernian Coast* on Coast Lines' Liverpool to London route along with her sister the *Caledonian Coast*, the latter never sailing on her intended east coast roster. The *Aberdonian* was sold for further service in the Far East.

*The last of the coastal liners, Coast Lines' **Hibernian Coast** (1947), was completed as the **Aberdonian Coast** for the Aberdeen Steam Navigation Company.*

(Coast Lines)

The last steamers on the Newcastle to London route of Tyne-Tees Shipping Company were the *Hadrian* and *Bernicia* built in 1923 by Swan Hunter & Wigham Richardson and Hawthorn Leslie & Company respectively. The new ships displaced six passenger and cargo steamers built in the 1900s, one of which, the *Richard Welford*, was broken up only in 1958. The *Hadrian* had a quadruple expansion engine, the *Bernicia* triple expansion, but both could maintain a service speed of 15 knots. They were not sisters in that the *Bernicia* had a gross tonnage of 1936 and accommodation for 136 first class passengers amidships and 100 second class in dormitory accommodation aft, and the *Hadrian* could carry 176 first and 508 second class passengers. The single fare for the 24 hour journey from the Tyne to London was 12/6d including berth and food.

A third ship, the *Alnwick*, was built by Swan Hunter in 1929 for the Rotterdam and Antwerp services. She deputised for the *Hadrian* and *Bernicia* during routine refits but had accommodation for only 75. In December 1932 the *Hadrian* and *Bernicia* were laid up as the Depression and competition from the railways took their toll. This fine pair of ships was sold to different Eastern Mediterranean operators in 1934. The *Alnwick* was then transferred to the London service, but rising costs and falling passenger numbers freed her for sale to Fred Olsen, for whom she became the *Bali* in 1935. The passenger service was discontinued altogether in 1936, having reverted in the final year to the charge of the Edwardian passenger cargo ships earlier displaced by the *Hadrian* and *Bernicia*.

*The **Bernicia** (1923) setting off on a coastal voyage with her deck cranes stowed.*

(Author's collection)

The convoy rescue ship

In September 1940, the Germans introduced wolf pack tactics, attacking convoys at night with a number of U-boats. They would sink a ship in an outside column and then, when an escort fell out to rescue survivors, slip into the convoy between the columns. At any time, half the escort could be engaged in rescue work and the other half hunting an attacker, leaving the convoy virtually unprotected. In January 1941, specially equipped merchant ships called convoy rescue ships were introduced, and one assigned wherever possible to each convoy. Typical convoy speed was about 9 knots, dictated, of course, by the slowest member of the fleet.

A total of 29 vessels were selected for the role, of which 27 were actually converted and two rejected as unsuitable (Table 3). Six of the ships were lost, the others remaining active until the end of hostilities in 1945.

The *Toward* and *Copeland* were two of a total of seven Clyde Shipping Company ships adapted for the role of convoy rescue ship in World War Two. The idea was that dedicated rescue ships attached to the larger convoys would both save lives as well as boost morale. It was agreed between the Admiralty and the Ministry of War Transport that relatively fast coastal ships be selected for conversion and that the coastal liner type vessel was ideally suited for the purpose given their low freeboard. The *Toward* and *Copeland* could maintain a good twelve knots, but if pushed could find a little more speed as required. Their role was specifically to rescue, accommodate and care for survivors.

Craig J M Carter, in an article that first appeared in *Sea Breezes* in January 1963 reporting the sale of the *Ulster Herdsman* for scrap, described her conversion to convoy rescue ship:

Bunker capacity had to be increased and as no cargo was carried, ballast had to be arranged. On top of the sand used for this purpose, hundreds of empty drums were placed to add buoyancy to the ship. Double tiered bunks replaced cattle fittings in the alleyways for at least 150 survivors, for whom messrooms, cooking and toilet facilities had to be provided. Full medical facilities, including operating theatre and sick bay, were under the direction of a Surgeon of the RNVR, and a Rescue Officer was in charge of the life saving gear, which included a fast motor boat.

The former *Copeland* and her consorts provided a valuable contribution to the war effort operating under difficult and often exposed conditions. Although several were lost at sea to enemy action none was released from duty until the cessation of hostilities. As for the *Toward*, she was operational from January 1941 and escorted 45 convoys and rescued 341 people. She was torpedoed and sank in February 1943 in Convoy ON 67 whilst returning from Halifax where she had earlier landed 164 survivors.

Table 3 : The Convoy Rescue Ships

Vessel	Year built	Owner	Passenger complement First/Second	Requisitioned	Number of rescues/ rescued	Lost or returned to owner
Aboyne	1937	Aberdeen, Newcastle & Hull Steam Co Ltd	12	June 1943	4/20	Returned June 1945
Accrington	1910	London & North Eastern Railway	150/300	March 1942	6/141	Returned May 1945
Beachy	1936	Clyde Shipping Company Ltd	42/40	October 1940	Unknown	Lost January 1941
Bury	1910	London & North Eastern Railway	150/300	August 1941	9/237	Returned June 1945
Copeland	1923	Clyde Shipping Company Ltd	60/30	December 1940	11/433	Returned June 1945
Dewsbury	1910	London & North Eastern Railway	150/300	July 1941	2/5	Returned June 1945
Dundee	1934	Dundee, Perth & London Shipping Co Ltd		April 1943	1/11	Returned June 1945
Eddystone	1927	Clyde Shipping Company Ltd	40/27	April 1943	2/64	Returned June 1945
Fastnet	1928	Clyde Shipping Company Ltd	40/20	June 1943	1/35	Returned June 1945
Goodwin	1917	Clyde Shipping Company Ltd	40/25	December 1942	5/133	Returned June 1945
Gothland	1932	Currie Line Ltd		November 1941	8/149	Returned June 1945
Melrose Abbey	1929	Hull & Netherlands Steamship Co Ltd	84/38	February 1941	5/85	Returned June 1945
Perth	1915	Dundee, Perth & London Shipping Co Ltd	60/72	October 1940	14/455	Returned June 1945
Rathlin	1936	Clyde Shipping Company Ltd	42/40	July 1941 [a]	13/634	Returned June 1945
St Clair	1937	North of Scotland, Orkney & Shetland St Nav	144/86	July 1940	0/0	Returned June 1945
St Sunniva	1931	North of Scotland, Orkney & Shetland St Nav	112/95	September 1942	0/0	Lost January 1943 [b]
Stockport	1911	London & North Eastern Railway	150/300	July 1941	10/413	Lost February 1943
Toward	1923	Clyde Shipping Company Ltd	60/25	December 1940	10/337	Lost February 1943
Zaafaran [c]	1920	Ministry of Transport	12	October 1940	7/220 [e]	Lost July 1942
Zamalek [d]	1921	Ministry of Transport	12	March 1940	19/617 [e]	Returned June 1945

Notes:

[a] Became a rescue ship only in July 1944
[b] Believed lost because of icing up when two days out of Halifax
[c] Formerly the *Philomel* of the General Steam Navigation Co Ltd and managed by GNSC for the Ministry of Transport
[d] Formerly the *Halcyon* of the General Steam Navigation Co Ltd and managed by GNSC for the Ministry of Transport
[e] Bought by the Ministry of Transport from Khedival Mail Lines, of Alexandria, which had acquired the ships from GSNC in 1934 (see Chapter 6)

In addition there was the Dutch-owned *Honestroom* which was released from duty in May 1941, the diesel-driven *Pinto* of Macandrews, Prince Line's *Syrian Prince* and the *Walmer Castle*, in addition to four corvettes. The *Pinto* and *Walmer Castle* were lost to enemy action.

Chapter 6 : Just a few passengers

Many of the overnight cargo and cattle boats around our shores used to have facilities to carry a few passengers. Typically twelve passengers were catered for, adverts sometimes stipulating male only passengers, usually with a small saloon below the bridge with adjacent cabins. Some could carry greater numbers and of these the last of the Holyhead to Dublin North Wall cattle boats with steam reciprocating engines (later ships were steam turbine powered and diesel) particularly fires the imagination. This was the *Slieve Donard*, delivered to the London, Midland & Scottish Railway in 1922 by Vickers Limited at Barrow. Her twin sets of four-cylinder triple expansion engines drove her across the Irish Sea at 16 knots.

*The **Slieve Donard** (1922) alongside at Holyhead and seen from the deck of a passing mail ship in September 1949.*
(Mike Walker collection)

The bill for the new ship was picked up by the American government. This was in recompense for the loss of the 1908-built *Slievebloom* (one word rather than two as often reported) which had been cut down and sunk by an American cruiser in 1918. This was also why the replacement had to be built to a design that was conceived in 1904, the new *Slieve Donard* being of the same basic design as the *Slievebloom*. The *Slievebloom* was herself the third ship in a class which had first emerged 18 years previously.

It should be remembered that the cargo and cattle service to the North Wall started in 1853 as there was no room to develop cargo facilities at the mail boat terminal in Kingstown (later Dun Laoghaire). The *Slieve Donard* continued the nomenclature based on Irish mountains (Slieve deriving from the Irish word Sliabh). The ship was of modest dimensions and designed primarily as a cattle carrier, she also had comfortable accommodation for 134 passengers. The reason for such a complement of passengers was that the *Slieve Donard* was responsible for the special Thursday daylight sailing from Dublin known as the 'Kings of the Cattle Market', on which the cattle dealers returning from market with their animals would receive a three course luncheon. The passenger facilities were withdrawn at the outbreak of war in 1939 when the extra pair of lifeboats on the poop were removed. The ship survived until withdrawn and scrapped in 1954.

*One of a pair of engines awaiting installation in the **Slieve Donard** (1922) and seen in the workshop of Vickers Armstrong during construction.*
(Barrow Museum Service)

More typical of this class of ship was the Clyde Shipping Company's trio *Tuskar* built at Dundee in 1920, *Skerries* dating from 1921 and *Rockabill* completed in 1931. These ships could each carry twelve passengers on services between Waterford and Liverpool and Waterford and Bristol. The passenger accommodation on the *Skerries* was withdrawn in the mid-1930s, but the other two continued to carry passengers. The *Tuskar* was a war loss - the Bristol passenger service was not reinstated after the war. The *Rockabill* continued on the Liverpool route carrying passengers to the last, providing a weekly service between there and Waterford. Replaced by a diesel driven cargo only vessel in 1962, the *Rockabill* was the last coal burning steamer in the Clyde Shipping Company fleet and the last to carry passengers. She was always noted for her high degree of maintenance, with shining paintwork and smooth running triple expansion machinery.

A number of railway cargo ships had facilities to carry about twelve passengers. A pair built for the Great Western Railway by Swan Hunter & Wigham Richardson at Newcastle were the *Sambur* and *Roebuck*. They were odd, rather stunted looking ships. This was because they had originally been designed with two holds aft of the central superstructure, but were built with only one to accommodate the restricted harbour at St Helier. They had twin sets of three-cylinder triple expansion engines which drove them at 13 knots. They served the Weymouth to Channel Isles routes throughout their careers apart from the war years when they assisted at Dunkerque, saw service as barrage balloon ships and helped construct the Mulberry Harbours. At the Weymouth station they were advertised with accommodation for ten male passengers, later reduced to eight saloon berths with one cabin.

Entry to the confines of the harbours at Weymouth, St Peter Port and St Helier normally required a stern first approach to the berth whilst paying out the anchor cable to reduce way on the vessel and later, on departure, to assist in pulling the vessel off the berth. These manoeuvres were accompanied by much hissing of steam from deck machinery and clanking of windlasses and chains. The ships remained in service until the mid-1960s.

*The **Roebuck** (1925) arriving at Weymouth stern first in July 1964. Note the use of the anchor to assist steerage and subsequent departure.*

*The **Selby** (1922) was built for Wilson's and North Eastern Railway Shipping Company and managed by Ellerman's Wilson Line until the formation of Associated Humber Lines in 1935.*

(Author's collection)

A number of other railway owned cargo steamers were equipped to carry up to twelve passengers. These included the London, Midland & Scottish Railway steamers **Dearne**, **Don**, **Hebble** and **Rye**. These ships became part of the newly-formed Associated Humber Lines in 1935. Most of the ships in the joint Wilson's and London & North Eastern Railway Company had first class accommodation for twelve passengers. The final pair built for this consortium were the **Selby**, which came from the John Duthie Torry Shipbuilding Company at Aberdeen in 1922, and the **Harrogate** which was delivered by Ramage & Ferguson at Leith in 1925, although neither was given passenger accommodation. They too became members of the Associated Humber Lines fleet in 1935, and both ships survived until 1958 when they were sold for demolition.

The Southern Railway commissioned a series of nine new twin screw steamers between 1925 and 1928. Of these, the **Haslemere** and **Fratton** along with the **Ringwood** could carry eight passengers, whereas the sisters **Whitstable** and **Hythe** could accommodate five and the **Deal** and **Tonbridge** none. They were fast for cargo ships, with a speed of 15 knots. The **Haslemere** and **Fratton** each cost £42 250 to build, some £3 000 more than the **Tonbridge**, the difference in price representing the addition of the cabin accommodation on the **Haslemere** and **Fratton**. Other ships in the class also carried some passengers. They were deployed on Dover Strait services and the Channel Island routes. All these ships were coal burners, and the last surviving member of the group was the **Deal**, which was sold for demolition only in 1963.

The tenth vessel in this class was to be named **Camberley**, but was delivered as the passenger and car carrier **Autocarrier**. She was completed in 1931 by D & W Henderson & Company at a cost of £49 150, and was slightly larger than the nine earlier cargo ships. She was different in that she had been adapted to carry between 30 and 35 cars, craned on and off, along with their owners. She had a certificate for 120 passengers on service between Calais and Dover, and was equipped with twin four-cylinder triple expansion units which provided a service speed of 15 knots.

And the reason why the Southern Railway was so keen to help car drivers who had shunned their railway services was competition. Trading under the name of Townsend Brothers, Mr Stuart Townsend, a keen motorist, had used the steam coaster **Artificer** to inaugurate a car-carrying service to France in 1928. Originally the new service was only intended to run for one month to get the Southern Railway to reduce their tariffs for cars to a reasonable level (£2 for a single crossing on the **Artificer** compared with nearly £6 on the railway ships). The service was such a success that it continued through the winter – a $2^1/_2$ hour journey with twelve deck passengers, the remainder transferred to the railway steamer.

The **Artificer** was replaced the following year by a larger steamer the **Royal Firth**. However, the arrival of the converted flush-decked minesweeper **Forde** in 1929 which offered space for 30 cars and 168 passengers (307 on a daylight Steam Limited certificate) at a speed of 13 knots was a landmark. She cost £5 000 to buy from the Admiralty and a further £14 000 to convert. Chartered cargo ships were also used on the service and the passenger accommodation was often full in summer. This was the beginning of Townsend Ferries, and was clearly of concern to the railway company. The **Forde** had coal-fired, Yarrow-type water tube boilers supplying steam to twin triple expansion engines.

Initially the Southern Railway had reduced the car tariff to £1 and dedicated one of the cargo fleet to carry unaccompanied cars. This was where the *Autocarrier* came in and from then on a competitive service was established on a more or less even footing. The *Forde* was used as a salvage ship in the war whereas the *Autocarrier* became a troop transport, attended at Dunkerque where she was one of the last ships to leave, and later acted as a recreation ship. Her final wartime duty was to assist in the repatriation of the Channel Islanders.

The *Forde* was displaced in 1950 by another converted naval ship, the *Halladale*, this time with turbine engines. The *Forde* ended her career on the Gibraltar to Tangier ferry route. The *Autocarrier* was withdrawn and scrapped in 1954, having spent her last seven years on a variety of routes including the Folkestone to Calais cargo service. She had been displaced as a car ferry by the post-war conversion of the turbine passenger steamer *Dinard* which had then taken up this role. The *Autocarrier* had previously also operated some relief services to the Channel Islands.

An interesting pair with day accommodation for 50 passengers were the *Minard* and *Ardyne* built in 1926 and 1928 respectively for the islands and coastal routes of Clyde Cargo Steamers Limited. They principally traded between Glasgow and Rothesay and Glasgow and Arran. In 1937 the company was merged into the Clyde & Campbeltown Shipping Company which became part of the British Transport Commission in 1949. In post-war years the ships were reduced to carrying only twelve passengers. Both ships were withdrawn and scrapped in 1955, with the arrival of the three ABC class vehicle ferries in 1954 for the Caledonian Steam Packet Company for service on the lower Clyde.

The London, Midland & Scottish vessels *Aire*, *Blyth* and *Calder*, dating from 1930 and 1931, each had accommodation for twelve passengers. These ships were oil burners. The *Calder* was lost on passage from Hamburg to Hull in 1932, and the *Aire* was sunk in collision outside Goole in 1958. The *Blyth* survived until 1959, the *Aire* and *Blyth* having been integrated into Associated Humber Lines services since 1935 but used for ferry duties to Orkney during part of the war.

The British & Continental Steamship Company traced its ancestry back to the Saint Georges Steam Packet Company formed in 1821. The new company was created in 1921 and retained the white and black funnel colours of the former Cork Steamship Company and incorporated the St George's cross in its house flag.

*The British & Continental Steam Packet Company's **Clangula** (1954) seen in the Manchester Ship Canal below Irlam Park Wharf in April 1963.*

Most of the ships had comfortable accommodation for up to twelve passengers. Round trip cruises from Manchester, Liverpool and Glasgow via Southampton on the weekly service to Amsterdam, Rotterdam and Dunkerque or Antwerp, Ghent and Terneuzen were extremely popular. Vessels engaged in these services which had passenger accommodation included the *Dafila* and *Tadorna*, both lost in the war, and the younger *Dotterel* and *Egret* which dated from 1936.

The occupation of the low countries in 1940 prevented the company's services from continuing and the ships were diverted to general duties. Services resumed after the war and a fleet of twelve steamers was again in use. However, by the mid-1950s the fleet was reduced to six, the *Egret* being displaced by a new motor ship of the same name in 1957, and the last of the pre-war steamers, the *Dotterel*, being sold for scrapping in 1961. In these latter years services were offered jointly with the Holland Steamship Company.

The only oil burners in the fleet were the post-war builds *Ardetta* and *Bittern* which came from the yard of Cammell Laird in 1949 and the *Clangula*, the last of the steamers to be built for the company, which came from Cammell Laird in 1954. Although all three had facilities to carry some passengers, the accommodation was little used and the service never attained the popularity with the travelling public that it had enjoyed in the 1930s. The company became wholly owned by its Dutch associates in the 1960s, and ceased trading in 1972 in response to competition from the shorter channel crossings and improved roads.

The Currie Line of Leith replaced a fleet of ageing North Sea packets with two handsome ships, the *Courland* and *Gothland* in 1932. They had a single tall funnel, cruiser stern and slight sheer with a balanced profile. The hulls, whilst all-rivetted, were strengthened for navigation in ice. They had a single triple expansion engine with steam provided by two single-ended return-tube boilers giving them a service speed of 14 knots. Luxury accommodation was provided for twelve passengers in five twin cabins on the Shelter Deck and two single cabins on the Boat Deck. The quality of the cabin accommodation was such that they were always referred to as staterooms. The dining saloon was panelled in Honduran mahogany and fitted with an open fireplace in true 1930s ocean liner style. Above, and connected by an elegant staircase, was the smoke room, panelled this time in Austrian oak and furnished with leather armchairs. The *Gothland* maintained her owner's connections between Leith and Hamburg and Leith and Copenhagen until she was sold in 1958, whilst the *Courland* had been an early war loss.

Currie Line also maintained the 1928-built *Hengist* and *Horsa* on the Copenhagen route throughout the 1930s. Only the *Horsa* survived the war and she terminated the regular Leith-Copenhagen passenger service when she was withdrawn in 1955, and was sold to William Sloan & Company to become *Endrick*. She was broken up in 1959.

Palgrave Murphy Limited of Dublin offered passenger accommodation aboard their ships travelling to Continental ports. Two of the favourites were the steamers *City of Amsterdam*, built in 1921 as the *Delgany* for the Dublin & Silloth Steamship Company, and the *City of Hamburg* which was built originally for the Aberdeen, Newcastle & Hull Steam Company as the *Aboyne* in 1937 (see also Chapter 5). These were sold in 1959 and 1958 respectively.

After the war they bought the twelve-passenger *Skerries* from the Clyde Shipping Company and renamed her *City of Waterford*, the former *Goodwin* which had become *North Tipperary* and was renamed *City of Cork*, and the General Steam Navigation Company's steamer *Seamew* which was renamed *City of Antwerp*. Of these the *City of Waterford* served only two years before she was sunk in collision off Beachy Head, the *City of Cork* was scrapped in 1953 and the *City of Antwerp* was sold in 1952. Passenger numbers were mostly confined to twelve under Palgrave Murphy ownership. With the sale of the *City of Amsterdam* in 1959 passenger services ceased although cargo services to the continent continued to operate with diesel vessels.

Passengers had been carried aboard the cargo ships of the General Steam Navigation Company since the early days. Passenger carrying was still significant on the eve of the First World War, but had shrunk to almost nothing by the outbreak of the Second. Whether this was the result of company policy or whether it was for operational reasons is not clear, but in 1913 the General Steam Navigation Company carried passengers on its services from London to Bordeaux, Edinburgh, Harlingen, Ostend and to Mediterranean ports including Genoa, Livorno, Naples, Messina, Catania and Palermo, as well as from Parkeston Quay to Hamburg. By 1939 only the London to Bordeaux and London to Leith routes still offered limited passenger accommodation. These services were not resumed after the war.

Some of the last of the passenger carriers to be built were the *Halcyon* and *Philomel*, which came from the Ailsa Shipbuilding Company in 1920 and 1921 respectively, and the *Heron* and *Starling* dating from 1920 and which were sold to the United Baltic Corporation in the 1930s (see Chapter 4). The *Halcyon* and *Philomel* had a gross tonnage of 1566, and had triple expansion engines supplied by twin single ended boilers and a service speed of 13 knots. They were designed for the Bordeaux service and offered comfortable accommodation for twelve passengers. However, they were found to be too expensive to survive the Depression on this route and were withdrawn and sold to Egyptian owners in 1934. They were bought back by the UK Ministry of Transport in 1940 to become the Convoy Rescue Ships *Zaafaran* and *Zamalek* and managed by GSNC (see Table 3). Commodore Wilson and Captain Birch, the former being honoured by the Bordeaux Chamber of Commerce for his services, closely supervised the ship's GSNC careers. Accommodation for four or six passengers was retained on a handful of other vessels during the 1930s. The Bordeaux service was taken over by the *Philomel*, built in 1927, and *Laverock*, built in 1909, although they were replaced in 1937 by motor ships.

MacAndrews and Company offered 16 and 15 day cruises to Spain in the 1930s at a cost of £1 per day. However, these were mainly offered by the newer diesel members of the fleet and only rarely by the older steamships.

A number of other companies offered limited and sometimes seasonal berths for passengers on their cargo routes, but few of these continued to do so after the Second World War. The Wilson Line was one of these.

A distinctive group of passenger carriers were the Wilson Line ships with a combination triple expansion engine and low pressure turbine, with double reduction gearing and hydraulic coupling. The first was the *Consuelo* completed in 1937 for their overseas liner service. For the North Sea routes the *Volo* and *Tasso* were completed in 1938 with accommodation for 12 passengers, the 1940-built *Ariosto* and *Angelo* and the post-war builds *Tasso*, a new *Ariosto* replacing her namesake which had been lost in the war, *Malmo*, a new *Volo*, *Tinto*, *Carlo*, *Truro*, *Leo* and *Silvio* which were all delivered by 1947 also had limited passenger accommodation. The Wilson Line's so-called 'Green Parrots' were characterised by their green hulls and red funnels. Cabins usually comprised five twin-berth and two single-berth cabins.

The service speed of the ships was 13½ knots and all were strengthened for navigation in ice. The ships could navigate slowly through relatively thin ice in which even a small pressure ridge would bring them to a jarring halt. As the ship backed slowly off the ice it would leave a line of red paint from the boot topping at the bottom of the V cut into the ice. Popping rivets and leaking plates were commonplace in the worst of the winter months! Reversing was difficult as the propeller was unprotected and care had to be taken to ensure no contact with the ice. In the early 1950s it was possible to do a ten day round trip to the Baltic from Hull for only £40, inclusive of meals, and reduced to only £20 in the winter months. By all accounts, the winter voyages offered the most passenger entertainment whilst navigating in ice! These ships were progressively withdrawn in the mid-1960s when the cost effectiveness of their engines became questionable. The last to be sold was the *Truro* which went to Kuwaiti owners in 1968.

*Typical of the Wilson Line "Green Parrots" was the **Volo** (1946) complete with ice-strengthened hull.*

(Author's collection)

Other ships of this type were the *Cicero* and *Rollo*, also capable of taking twelve berthed passengers, but with superior quality accommodation, and delivered from the Henry Robb yard at Leith in 1954, and the *Borodino*, which had been delivered by the Ailsa Shipbuilding Company at Troon in 1950. The latter ship was designed for the Hull to Copenhagen route and could accommodate 36 passengers in first class and twenty in third class. Alas, her triple expansion combination low pressure turbines were expensive to maintain in an increasingly diesel era and the ship was withdrawn and scrapped at the tender age of only 17. The younger ships survived until Autumn 1970 when they were sold to overseas owners. Latterly they had been transferred from Scandinavian routes to the Wilson Line Mediterranean service based at London.

Other steam driven vessels equipped to carry twelve berthed passengers which deserve mention are Frank Bustard's Empire-class roll on roll off accompanied freight vehicle ferries, pioneers of the modern day freight ferry. These were built towards the end of the war as Type 3 Landing Ships Tank in Canada and the UK. Bustard acquired three of them from the Ministry of War Transport in 1946, initially at a daily charter rate of £13 each. Slightly modified, these twin screw triple expansion engined vessels inaugurated a service between Tilbury and Rotterdam and shortly afterwards between Preston and Larne, later also Belfast. The original trio was the *Empire Baltic*, formerly *LST 3519*, *Empire Cedric*, formerly *LST 3534* and *Empire Celtic* which had been *LST 3512*. The service speed of the ships was a leisurely 10 knots.

Additional former Landing Ships Tank were acquired and modified for the burgeoning freight services and the last of the steamers to remain operational before being displaced by purpose-built tonnage was the *Empire Nordic*. She had joined the fleet in 1955 and was sold for scrapping in 1966, the very last of the twelve passenger triple expansion engined vessels to fly the Red Ensign.

One of the ships managed for the Ministry of Transport by the Atlantic Steam Navigation Company, the *Empire Shearwater*, was chartered briefly to European Ferries (Townsend Car Ferries) during 1958 for use between Dover and Calais. However, the concept of heavy wheeled freight traffic on the Dover Strait was years ahead of its time and the project sadly failed.

*The **Empire Nordic** (1945) seen on 18 August 1966 in the River Ribble heading towards Preston Dock in her last year of service.*

The Clyde Shipping Company

Although better known for its tugs and coastal liners, the Clyde Shipping Company also operated numerous cargo vessels over the years, some with accommodation for a few passengers. The company dated from 1815 when two steam passenger luggage boats built in 1814 were acquired by a group of merchants to run between Greenock and Port Glasgow up the shallow and winding River Clyde to the Broomielaw. In 1856 the company was put up for sale including five tugs, three luggage steamers and eight lighters.

The new owners disposed of the luggage boats and lighters in the face of competition from the newly built Glasgow & Greenock Railway and the Clyde Trust's dredged and straightened channel to Glasgow. The company concentrated initially on Clyde towage, although former employee James Steel set up in competition on the Clyde to form rival towage company Steel and Bennie. Two cargo ships were bought in 1856 to inaugurate a service between Cork, Waterford and Glasgow. The lighthouse nomenclature for the cargo steamers was adopted in 1860 when the first *Tuskar* was delivered, whilst new tugs were given the famous Flying names with the delivery of the *Flying Childers* in 1856. By acquisition, the cargo route was extended to included Belfast in 1871 and Plymouth and Southampton in 1872. By 1884 the network included London and numerous additional ports, with many of the steamers offering limited accommodation for a few passengers and in due course developing the passenger facilities to offer accommodation in two classes.

In 1885 the Queenstown Towing Company (Queenstown now known as Cobh in Co Cork) was purchased and with it came the first tug/tender, the *Mount Etna*, which was renamed *Flying Eagle*. The company also took delivery of a couple of cargo ships for the international tramp trades in the late 1880s. The Waterford Steamship Company was acquired in 1912 giving access to Liverpool and Bristol. At the start of the Great War the company offered the following departures from Glasgow:

Monday: To Waterford and Plymouth.
Tuesday: Belfast, Southampton, Newhaven and Dover, returning via Plymouth and Belfast.
Wednesday: To London and back via Belfast.
Thursday: To Waterford, Southampton, Newhaven, Dover and London, returning via Southampton, Plymouth and Waterford.
Friday: To Belfast, Plymouth and London returning via Belfast.
Saturday: To London and back.

After the war the foreign going services were abandoned, but five new steamers were added for the coasting routes to replace war losses: *Tuskar*, *Aranmore*, *Skerries*, *Toward* and *Copeland*.

Competition with the shorter east coast Leith to London routes offered by the London & Edinburgh Shipping Company, the General Steam Navigation Company and the Carron Company led to a cargo pricing conference which in 1946 was formalised as London, Scottish Lines. Post-war rebuilds were cargo only diesel ships, but the *Rathlin* continued to offer passenger accommodation until 1953. The last steamer in the company was the Liverpool to Waterford steamer *Rockabill*. Thereafter the company retained diverse interests centred principally on towage and ceased to trade in 2000.

Chapter 7 : To the islands

A number of very fine steam reciprocating engined passenger and cargo ships serviced the needs of islanders and the remoter parts of the coast that were not adequately served by overland transport. The ships of the North of Scotland, Orkney & Shetland Shipping Company and of the Isles of Scilly Steamship Company include some particularly fine examples, some of which remained in service well into the diesel era. By way of contrast, the Western Isles of Scotland were largely serviced by diesel ships from the 1930s onwards and steamers were phased out relatively early. On the more sheltered routes such as the Thames, Clyde, Forth and Bristol Channel, paddle steamers ruled the roost; their shallow draft and exceptional manoeuvrability being ideally suited to working shallow tidal piers. Turbine steamers were also in use on the Clyde and Thames although direct-drive steamers were preferred to the more sophisticated geared turbines because the momentum of the gear trains inhibited the rate at which the ships could be slowed down on approach to the piers. They also drew considerably more water than the paddle steamer.

That being so, there was one most interesting steam reciprocating engined ship operating an island type service in estuarial waters. This was the *Dalriada* built by Robert Duncan & Company at Port Glasgow, at a cost of £42 500, and delivered to the Campbeltown and Glasgow Steam Packet Joint Stock Company Limited in 1926 to run alongside the elderly *Davaar*. The *Dalriada* was one of the fastest single screw steamers ever built and her four-cylinder triple expansion engine maintained a service speed of 17$^{1}/_{2}$ knots. She spent her time running between Glasgow and Campbeltown calling at Lochranza, Pirnmill, Carradale and Saddell. On Saturdays she did a double run from Greenock offering excursionists an afternoon at sea, and ran occasional cruises from Campbeltown. On a Steam 5 Certificate she could carry 1294 excursionists. Her running mate, the *Davaar*, dating from 1885, sported a handsome clipper-shaped bow.

The *Dalriada* carried the traditional company funnel colours of black with a broad red band. This became red with a black top in 1937 when the company was bought by Clyde Cargo Steamers Limited, to become part of the small fleet of the MacBrayne subsidiary which then became the Clyde & Campbeltown Shipping Company. The *Davaar* was taken south during the war for possible use as a blockship to be sunk at Newhaven, but was broken up on the beach in 1943, and the *Dalriada* was lost in the Thames estuary in June 1942 whilst working as a salvage ship.

*A magnificent view of the **Dalriada** (1926) reproduced from a postcard by Roberston & Company, of Gourock.*
(Author's collection)

Two venerable Victorian steamers on the Clyde to the Western Isles passenger and cargo service of John McCallum & Company and William Lang – Martin Orme (later McCallum, Orme & Company) were the *Hebrides* and the *Dunara Castle*. The *Hebrides* was built in 1898 by the Ailsa Shipbuilding Company and had a gross tonnage of 585. Her triple expansion machinery gave her a speed of almost 13 knots. She could carry 50 passengers and in summer carried out extended cruises to St Kilda and Lewis. She was given a new boiler and furnaces in 1937. Passenger carrying ceased in 1939, and the vessel became part of the David MacBrayne's fleet until withdrawn and scrapped in 1955. The *Dunara Castle* led a similar career, but starting 23 years earlier when she was delivered from Messrs Blackwood & Gordon at Port Glasgow. Slightly smaller, she had a gross tonnage of 423 when built and could also berth 50 passengers. She served the Glasgow and West Highlands passenger and cargo route for 73 years and was sold for demolition shortly after entering the David MacBrayne fleet in 1948.

Duckworth and Langmuir, in *West Highland Steamers*, reported:

Fitted for a time with two funnels, the *Dunara Castle* was reboiled in 1882 (when the machinery was compounded), and again in 1894, from which date she was single funnelled. From time to time various improvements were made, such as the installation of electric light, wireless, etc.; and a small upper deck aft, was added about 1945. She was a handsome ship and a fine sea boat.

The last passenger steamship to be built for David MacBrayne Limited was the *Lochness*, a replacement for the *Sheila* which was wrecked at Loch Torridon on New Year's Day 1927. The *Lochness* was built by Harland & Wolff at Govan with twin triple expansion engines and two single ended oil-fired Scotch boilers. Her service speed was 14 knots. She was designed for the hard working Mallaig and Kyle of Lochalsh to Stornoway route, which only allowed her a Sunday lay over for maintenance. A utilitarian but imposing looking vessel she had comfortable first class accommodation as well as third class, the latter occasionally the subject of criticism in the Stornoway Gazette. Displaced in 1947 by a new build, the *Lochness* saw duty on a variety of other routes based at Oban although she had become expensive in both fuel and crewing. As spare ship, she was sold in 1955 to begin a new career in the Greek Islands, being withdrawn and scrapped only in 1974. The very last steamer in the fleet was the small 'Empire' Canadian war-built *Loch Frisa*, sister of *Lochbroom*, the former only being displaced in 1963 from the West Highland cargo service.

*The **Lochness** seen dressed overall early in her career. She had already lost her grey hull in favour of the traditional and more visible black.*

(Donald Meek collection)

It is worth reflecting that the island steamers were core to the life of the island communities that they served. Professor Donald Meek at Edinburgh University explains that the steamer and her diesel successors were the island's only outlook on the world. They allowed him as a young boy to visit eye specialists in Glasgow, who repaired his defective vision. The steamer also gave him access to Oban High School. Without the ship, his home on the island of Tiree was non-viable; the steamer was the only contact that his boyhood world called Tiree had with the greater world that started at the quayside at Oban.

MacBrayne's steamer *Claymore* was the mainstay of the year-round Glasgow to Stornoway passenger and cargo route from her delivery in 1881 until withdrawn in 1931. Throughout her career she was driven by a two-cylinder compound engine fed by two single-ended Scotch boilers. The last steamer to be built for the Glasgow to Stornoway seasonal cargo and passenger cruise ferry service was the stately *Chieftain*, spending most of her winter months in mothballs. She was delivered from the Ailsa Company's yard at Ayr in 1907 resplendent with clipper bow and a single triple expansion engine and modern comfortable accommodation. Her service speed was 14 knots. The first class saloon was forward on the Shelter Deck and occupied the full width of the ship. Cattle were carried in the 'tween decks. Alas, a declining travelling public and the ravages of the Great War allowed the sale of the ship to the North of Scotland, Orkney & Shetland Steam Navigation Company in 1919, to become their *St Margaret*. Given alterations to her upper structure by her new owners, she gained a propensity to roll quite severely. Contrary to popular belief, her deck machinery was steam driven and not electrical – there was only the lighting circuit which was a low voltage DC system of modest power.

She lasted in this guise until 1925 when she was again sold to follow a nomadic career on both sides of the Atlantic. She was hindered on the North Company routes by the lack of a well deck to assist cargo handling. She was eventually broken up in 1952.

The *St Margaret* was replaced by a new build, the *St Magnus*, which came out in 1924. She was a comparatively large ship for the company with berths for 234 in first class and 84 in second class, the latter arranged in two 42 berth dormitories. The first class public rooms also had settees that converted into an upper and lower bunk providing additional peak capacity. She was also exceptional in that she had two holds forward instead of just one, and the normal single hold aft. The ship serviced the intermediate route to the islands until opportunity was taken in 1956 to convert her from coal to oil fuel and to upgrade the dormitory style accommodation. The *St Magnus* retained 'tween deck cabins that could be removed in winter to accommodate cattle stalls, the crew of course always referring to these cabins as the cattle stalls to the consternation of passengers allocated these berths. She was finally withdrawn for scrap in 1960 with the arrival of the new diesel-driven *St Clair*.

*The **St Magnus** (1924) served the 'North Company' until 1960.*

(P&O Group)

The next steamer to be built for the company was the magnificent *St Sunniva* which entered service in June 1931. Considering that the company had commissioned conventional straight stemmed vessels for several decades previously, it was perhaps surprising to see that the new ship had classic Victorian lines and a clipper bow. For that she was welcomed as a striking vessel, pleasing to the eye, and furthermore, one that was found to be efficient and popular with travellers. The *St Sunniva* had a slightly longer hull than the *St Magnus*, but the central accommodation structure was smaller and her gross tonnage, at 1368, was consequently also smaller. She was a single screw vessel with steam supplied by two coal-fired Scotch boilers to a single triple expansion engine. Her service speed was 15 knots.

The *St Sunniva* had only 112 first class berths, ten in 2-berth cabins, 14 in 4-berth and a ladies' dormitory cabin for sixteen. Fifty four could be accommodated in the dining saloon. Second class included a men's dormitory with 54 berths and settee berths for up to 18 in the public rooms. The ladies' cabin could sleep 23, strangely the only access to this cabin was via the men's dormitory; 380 passengers could be carried in total. There were two holds, one forward and one aft. The new ship served the direct route between Leith, Aberdeen and Lerwick until the outbreak of war. For the eight winter months she would lay up in the Victoria Dock at Aberdeen, the winter ship being the former *Lairdsbank*, built in 1893 as the Laird Line's *Olive* (see Chapter 4) now the *St Catherine*, which in turn took her own place in Victoria Dock for the summer months. The elderly *St Catherine* was withdrawn and scrapped in 1937 and thereafter the winter service fell to the *St Magnus*, requiring the *St Sunniva* to extend her season by two months.

The *St Sunniva* was a popular ship and occasionally had to leave intending passengers on the quayside. Her schedule required a Monday morning from Leith with a seven hour voyage up to Aberdeen, sailing again after a few hours on Monday evening for the 16 hour passage to Lerwick. From Tuesday afternoon to departure for the return sailing at midday on Wednesday she worked cargo. She spent Thursday morning at Aberdeen leaving for Lerwick again in the afternoon. She would then sail at 5 pm from Lerwick once more for Aberdeen, and on to Leith, arriving early Sunday evening. Each visit to Lerwick coincided with outward and inward connections to the outer isles.

This happy existence came to an end in August 1939 when she was requisitioned for duty at Scapa as a guardship. Both the *St Sunniva* and the *St Magnus* were active in the Norwegian Campaign in 1940 and both received the Battle Honour Norway 1940. Returning to guardship, the *St Sunniva* was then converted to convoy rescue duties in 1942 (see Chapter 5) and put under the management of the General Steam Navigation Company. She had a complement of 32 men in the deck and engine room departments, 12 in the gun crew, nine in the catering department, eight in signals and three medical staff. Her first transatlantic convoy was to Halifax, Nova Scotia in January 1943. Worsening weather and bad icing conditions saw the convoy broken up. The *St Sunniva* was last seen two days out of Halifax, it being assumed that build up of ice destabilised the ship causing a sudden capsize before the alarm could be raised. A tragic end to a fine ship. Meanwhile one of the Canadian escorts returned to Halifax with a build up of over 3 metres of ice on the bridge front at deck level, tapering to one metre beneath the bridge wings.

The next steamer to be built for the company in the 1930s was an equally fine ship, the *St Clair*. She was delivered from the Aberdeen yard of Hall Russell and Company in April 1937. The *St Clair* was the very last conventional or classic overnight passenger and cargo ferry to be equipped with a single triple expansion engine and single screw under the Red Ensign. The new ship differed fundamentally from her predecessors in two ways. Firstly the superstructure was entirely plated in beneath the Bridge and Boat decks so that the traditional deck houses were encompassed in the main structure apart from the second class deck house aft. Secondly, the passenger layout was reversed so that first class passengers now occupied the forward section and second class passengers the after section of the accommodation.

*The very last classic passenger and cargo coastal steamer to be commissioned was the North Company's **St Clair** (1937).*
(Author's collection)

The first class lounge and observation area with its large windows faced forward on the Shelter Deck. Beneath this on the Main Deck was the dining saloon, at last referred to as the restaurant, with seating for 56. The first class cabins were on the Boat Deck, and comprised 36 convertible 2 and 4 berths. There was a small bar on the same deck. There was no facility for additional overflow berths in the public rooms. The second class smoking room was in the after end of the stern deckhouse, where there was also four passenger cabins. The saloon was on the Main Deck along with the remainder of the second class cabins. A total of 86 second class berths could be provided.

The *St Clair* commenced service in May 1937, working largely from Leith to Lerwick but operating the tourist 'west coast service' from Aberdeen to Stromness in Orkney and Scalloway and Hillswick in the Shetlands. During the war she operated as *HMS Baldur*, taking part in the Occupation of Iceland and later reverting to her own name as a convoy rescue ship. She again took up civilian duties in the summer of 1945 and adopted the Monday and Thursday evening departures from Aberdeen direct to Lerwick.

In 1959 a new diesel-powered *St Clair* was under construction (surprisingly also single screw), and the old ship was firstly renamed *St Clair II*, then *St Magnus* and finally *St Magnus II*. It was under this name that she finally sailed to the breaker's yard in 1967 having given her owners 30 years of solid service. A fondly remembered ship and one which had been very much the last of the steamers.

The outer isles and inter-isles service in Shetland were provided by the steamer *Earl of Zetland*. She had been built for the Shetland Isles Steam Navigation Company in 1877 which became part of the North Company fleet proper in 1890. A new *Earl of Zetland*, the first diesel unit in the fleet, replaced the elderly steamer in 1939. In the Orkneys, the Pentland Firth crossing was maintained by the *St Ola*, built in 1892 and, like the old *Earl of Zetland*, she had a very long and successful career, being replaced in 1951 by a new diesel-driven *St Ola*. Only one of the North of Scotland, Orkney & Shetland cargo steamers had berths for a few passengers, although the post war diesel cargo ships had accommodation for twelve. The one steamer was the *St Clement*, built in 1928 with twelve berths in the midships superstructure. Her accommodation was particularly useful in coping with overflow traffic from the passenger steamers. The ship was lost in April 1941.

The Orkney North Isles services were the province of the Orkney Steam Navigation Company, formed in 1868, which latterly operated the *Earl Thorfinn* dating from 1928 and the *Earl Sigurd* built in 1931. These ships replaced the small *Countess of Bantry* which was originally built for service in Ireland. The *Earl Thorfinn* had a slightly larger and broader hull form than the *Earl Sigurd* and had a gross tonnage of 345 compared with her smaller sister's 221. The service speed was a stately 8 knots, sustained by coal-fired triple expansion engines. The *Earl Thorfinn* was retired in 1962 when the *Orcadia* was delivered to the Secretary of State for Scotland and managed by a newly formed and partly state owned Orkney Isles Shipping Company. The *Earl Sigurd* continued in service until 1969 to become the very last coal-fired passenger and cargo steamer in commercial service in UK waters.

*The **Earl Sigurd** (1931) leaving Kirkwall on a routine trip to Orkney on 21 August 1968.*

The last of the Isle of Man Steam Packet Company's passenger vessels to be built for the company with steam reciprocating engines (others were bought second-hand) was the *Snaefell*, which was delivered by Cammell Laird at Birkenhead in 1910. The company already had two turbine steamers, the *Viking* and the *Ben-my-Chree* but wanted an intermediate class vessel to maintain the winter Liverpool to Douglas route and the secondary Douglas and Ramsey routes in the summer. Her machinery was innovative comprising two sets of vertical 4-cylinder triple expansion engines which at 4300 indicated horse power gave the ship a service speed of 19 knots. She was modest in size being measured at 1368 tons gross but nevertheless could carry 1241 passengers and a crew of 43. She became an armed patrol vessel in the Great War and was lost in this capacity in June 1918.

The Snaefell (1910) seen in wartime camouflage livery.
(K E Abraham collection)

There was one cargo unit in the Isle of Man Steam Packet Company that offered accommodation for up to twelve passengers. This was the single screw triple expansion engined *Peveril* built at Cammell Laird in 1929. Her passenger accommodation was in great demand when she was operating the Liverpool to Ramsey cargo service. An oil burner, she was replaced by a motor ship of the same name in 1964 and sold for demolition.

The Peveril (1929) served between Liverpool and Douglas or Ramsey for 35 years.
(Richard Danielson collection)

The Scillonian (1926) served the Isles of Scilly until replaced by a diesel ship of the same name in 1956.
(Author's collection)

Finally, the last steamer of the Isles of Scilly Steamship Company deserves mention. Created in 1920, this company originally had the former coast guard cutter *Peninnis* on the route between Penzance and St Mary's. The route had previously been under the charge of various vessels. The *Lyonesse*, built in 1889 for the West Cornwall Steamship Company, with twin funnels and triple expansion engines, was by far the best known. The *Peninnis* was replaced in 1926 by a purpose built steamer, the *Scillonian*, which came from the Ailsa Shipbuilding Company at Troon complete with a single three-cylinder triple expansion engine. Capable of 12 knots, she was a lively sea boat given the shallow draught needed to access St Mary's Harbour on the islands. She had ample cargo space and had her own derricks and steam winches to work cargo, but also had accommodation for 390 day passengers. The *Scillonian* became a very popular ship, surviving the war years on the Scilly Islands run. She continued until 1955 when she was renamed *Peninnis* in anticipation of a new motor ship which was also to be called *Scillonian*, and which eventually displaced the old steamer in the spring of 1956. The old *Scillonian* had suffered three minor groundings during her long career. Otherwise she enjoyed an uneventful life, and was a most reliable unit that must have repaid her capital outlay time and again.

The North of Scotland, Orkney & Shetland Steam Navigation Company

The North Company can trace its ancestry back to 1790. In that year the Leith & Clyde Shipping Company was formed to provide a service between Edinburgh and Glasgow via the Pentland Firth (the Caledonian Canal was only opened in 1822). Using a fleet of small sailing vessels the company quickly established its reputation at numerous intermediate ports. It amalgamated with the Aberdeen, Dundee & Leith Shipping Company in 1820 to become the Aberdeen, Leith, Clyde & Tay Shipping Company. The word Tay was dropped from the title in 1824. The company operations were centred on Aberdeen and this is where it was administered from thereafter.

The Aberdeen, Leith and Clyde Shipping Company acquired its first steamer in 1821. This was the ***Velocity***, of 121 tons gross and built by Denny at Dumbarton. The order was precipitated by competition from a rival company, the Leith & Aberdeen Steam Yacht Company which had put a steamer on the Leith to Aberdeen route earlier that year. This was the only time the North Company had any serious competition in its long existence, and that ceased in 1826 when the rival ship had to be sold when the respective owning partners' agreement ceased. By that time the summer only steamship services had been extended to Inverness and finally as far as Wick.

In 1836 a new ship, the ***Soveriegn***, inaugurated a weekly service between Newhaven (adjacent to Leith), Aberdeen, Wick and Kirkwall, and extended to Lerwick initially on alternate weeks. Two years later the company won the prestigious mail contract to the islands. The company used the monarchy nomenclature for the next three decades, and the first screw steamer to be built for the company was the ***Queen*** in 1861. The first to adopt the famous saint nomenclature was the former paddle steamer ***Waverley***, built for the North British Railway in 1864, and acquired by the Aberdeen, Leith and Clyde Shipping Company in 1867 when she became the ***St Magnus***.

In June 1875 the company was reconstituted as the North of Scotland, Orkney & Shetland Steam Navigation Company. The title was changed in 1953 to North of Scotland, Orkney & Shetland Shipping Company. In 1961 the company sold out to the Coast Lines Group for a surprisingly small price, it being said at the time that this reflected the North Company's reluctance to consider roll on roll off traffic.

The identity of the North Company and its ships remained unchanged until the corporate image of parent P&O was installed on the fleet in 1975 after which the trading name became P&O Ferries (Orkney & Shetland Services) and the ships were reliveried in P&O colours. P&O withdrew from the route in 2002 having lost a competitive tender for the government subsidies to a new company, Northlink, which is jointly owned by the Royal Bank of Scotland and Caledonian MacBrayne. The simple plain dull yellow funnels and black hulls of the North Company will long be remembered, especially by the islanders.

Chapter 8 : Of tugs, tenders and excursion ships

Excursion ships, other than small river and harbour trip steamers, were not commonly equipped with steam reciprocating machinery and screw propulsion. There were, however, several built for service in the Fylde and Morecambe Bay area including the ***Deerhound*** and the ***Robina***. The ***Deerhound*** served for only four years before being sold and the ***Robina***, which was built in 1914, had little opportunity to demonstrate her worthiness, later leading a nomadic life and ending up in the Red Funnel fleet at Southampton. The ***Robina*** was a twin screw steamer of 306 tons gross and was eventually broken up in 1953.

*The nomadic excursion steamer **Robina** served in a variety of guises for a variety of owners during a 40-year career.*

(K E Abraham collection)

*The paddle steamer was the preferred model of excursion steamer until the late 1940s. The **Crested Eagle** (1925) of the General Steam Navigation Company was the first designed to burn fuel oil.*

(Author's collection)

The only other notable screw steamer built principally for excursion work was the Galloway Saloon Steam Packet Company steamer ***Roslin Castle***. Built at Leith for £12 200 in 1906, she quickly became popular on the Firth of Forth services and excursions until she was bought by the Admiralty for use as a tender in the First World War. The reason why an operator whose entire fleet consisted of paddle steamers should resort at that time to a screw propelled steamer was, coincidentally, the arrival of the turbine steamer. The prototype turbine steamer ***King Edward*** had been operating on the Clyde since 1901 and Galloway was enthralled by the idea of bringing a turbine steamer to the Forth. Financial considerations induced by the parent North British Railway restrained this vision and he had to make do with a screw steamer with conventional triple expansion engines. The ***Roslin Castle*** was nevertheless a much loved member of the fleet, with superior accommodation to any other steamer seen at that time on the Forth.

The Liverpool & North Wales Steamship Company bought a second hand steamer in 1922. This was the former coal-fired German tender ***Hörnum***, built as a minesweeper but never completed as such. She was very much a stop-gap to maintain excursions from Llandudno and was sold for further service in Italy five years later. This was the only screw steam reciprocating ship the company ever operated, other vessels being paddle steamers, the 1914-built turbine steamer ***St Seiriol*** which was requisitioned by the Admiralty and never used commercially, and the famous turbine steamers ***St Tudno*** and ***St Seiriol***, and the smaller motor vessel ***St Trillo***, which survived until 1962 when the company was wound up. The ***Hörnum*** was renamed ***St Elian*** for service at Llandudno. With twin fast-running triple expansion engines and water tube boilers working under forced draught, this handsome steamer could manage 15 knots.

She could accommodate 528 passengers and was used on a variety of routes. Her week under the command of Captain Cullen comprised Llandudno to Douglas on Tuesdays and Thursdays, Llandudno to Liverpool on Wednesdays, Liverpool to Menai on Fridays, and Liverpool round Anglesey most Sundays. In addition she sailed Liverpool to Llandudno on Monday afternoons and Liverpool to Menai on Saturday afternoons. She also did the 'Trip of the Season' from Liverpool to Bardsey Island. She was sold in 1927 and under Italian patronage she was converted to oil fuel and considerably rebuilt in 1947. Only withdrawn from service in 1977, she then adopted a static role at Salerno.

A number of screw steamers built for estuarine ferry work became excursion ships towards the end of their lives. Notable is the Mersey ferry *Royal Daffodil*, built for the county Borough of Wallasey in 1906 and which became the London Docks cruise excursion steamer for the New Medway Steam Company in 1933. She had been given royal status by King George V for her war service along with her sister *Iris*. It has been said on a number of occasions that the Borough of Wallasey showed lack of foresight in not insisting that the new owners changed her name, but in truth the majority opinion was very much against the name being changed on sale. Indeed, when challenged whether the New Medway Steam Packet Company 'Queen' nomenclature was to be applied, their Chairman stated:

You may rest assured the name *Royal Daffodil* will not be altered, on the contrary we shall treasure it and the ship.

Giving newly-built motorships the Royal names in later years was, however, an altogether different matter.

Another ferry that became an excursion steamer was the former Tilbury to Gravesend ferry *Gertrude* which had become the *Rochester Queen* in time for the 1933 season to run the Strood, Chatham and Sheerness to Southend service before being resold in the Autumn to become a tender at Gibraltar. Her main point of interest was that she was powered by two sets of vertical triple expansion engines. Six Mersey ferries (see also Chapter 9) found their way to Cork to become tenders and part time excursion ships:

An Saorstat, ex *Rose*-1927, built 1900, scrapped 1951;
Failte, ex *Lily*-1927, built 1901, wrecked 1943;
Blarney, used at Dublin 1932 until 1937, ex *Iris*, ex *Royal Iris*, built 1906, scrapped 1961;
Shandon, used at Dublin 1932 until 1939, ex *John Joyce*, built 1910, scrapped 1962;
Killarney, ex *Frances Storey*-1951, built 1922, scrapped 1962;
Bidston, built 1933, on charter for use in Cork in 1960s.

*The Blackpool Passenger Steamboat Company's **Minden** (1903) in service in the mid-1930s.*

(Author's collection)

The Mersey ferries *Bidston* of 1903 and *Upton* of 1925 both became excursion ships, the former as the *Minden* for the Blackpool Passenger Steamboat Company between 1933 and 1937 and the latter for Red Funnel at Southampton (without change of name) between 1946 and 1951. The *Upton* was placed on the Swanage to Bournemouth service and later the Southampton to Ryde route. It was found that the old Mersey ferry was difficult to steam on these longer routes and difficult to handle at piers without the benefit of a fast running tide and she was withdrawn at the earliest opportunity. The *Snowdrop* of 1910 was sold to the London & North Eastern Railway in 1936 to become the Granton to Burntisland ferry *Thane of Fife*. The service was suspended in 1940 when she was used as a tender until sold for demolition after the war.

Unlike excursion ships, tugs and tenders need bollard pull and screw steamers with reciprocating engines were an ideal configuration. There was a great number of them. In 1955, for example, the Admiralty operated 101 tugs in home waters: 10 were diesel and of the others, eight were steam engined paddle tugs and the remainder steam screw tugs including the wartime TID class of harbour tug. Many of these vessels had very long careers, the eldest ship in the fleet being the Portsmouth-based paddle tug *Volatile*, which dated from 1900. On the Thames, the entire fleet bar one of Sun Tugs were steamers, and all of William Watkins fleet were steamers. Diesel engines were more popular then with the operators of smaller tugs working in the docks and those used for lighter towage duties. At Liverpool, in the Bristol Channel and at Southampton, the entire Alexandra Towing fleet were steamers as were the Southampton-based Red Funnel tugs and the Lamey and Cock Tugs at Liverpool. The steam tug was still very much in evidence and only began to be phased out with new building from the early-1960s onwards.

By 1965 over half the Sun Tugs' fleet and that of William Watkins were diesel. Alexandra Towing were still 66% steamers and their entire fleet at Swansea was steam driven. The Red Funnel Group was more diesel than steam (see Appendix 1). Ten year later, in 1975, steam tugs were the exception. Then there were only nine operational steam tugs, and some of these were only rarely in steam. Three of the steamers were in the fleet of Falmouth Towage Company. However, there were also 36 former steam tugs which had recently been re-engined with diesel units. The steam tug was almost extinct.

Ordinary towage tugs had often been used in the past on seasonal excursions to boost their owners' revenues. The last tug in regular excursion use was the United Towing Company's *Yorkshireman*, built in 1928 for towage duties on the Humber, but based at Bridlington each summer with accommodation offered in two classes, complete with musicians to entertain, and a tiny bar. Her last season at Bridlington was 1955 and she was sold for breaking up ten years later. George Dickinson described the tug in an article which first appeared in *Sea Breezes* in 1965:

> There were two classes of passengers, the first having the forward and aft parts of the Bridge Deck for their exclusive use. The forward part of this deck was glassed in for the season with screens to form a miniature Promenade Deck. In autumn, when the tug returned to the Humber to resume her towing duties, the screens and the canvas used to cover the rails on this deck forward and aft were removed.

There were still a number of steam tug-tenders in operation in 1955, serving the dual needs of towing duties and servicing visiting liners anchored off shore. Tug tenders principally operated at Southampton, the Clyde and at Liverpool. These included the *Egerton* and *Flying Breeze* built in 1911 and 1913 respectively for service in the Alexandra Towing fleet at Liverpool, and the *Romsey* which was built in 1930 for use at Southampton. The last named replaced a tender of the same name, originally built in 1918 for the Admiralty, which was sold in 1929 to the French Line for whom she served another 30 years, despite being scuttled by the Germans at St Malo in the war. The greatest claim to fame of the new *Romsey* was carrying Winston Churchill out to *HMS Renown* on the Clyde in August 1943, despite the tender having been run down and sunk by the Burns & Laird Lines' *Lairdsburn* only the previous year!

The Southampton, Isle of Wight and South of England Royal Mail Steam Packet Company (Red Funnel) also operated the elderly *Paladin*. She was originally built for the Anchor Line to service the company's liner calls at the Tail of the Bank off Greenock in 1913, and was bought by the Clyde Shipping Company in 1939. The Clyde Shipping Company had earlier built and operated a pair of paddle tenders at Queenstown (Cobh). Aptly named *America* and *Ireland*, they served liner calls at Queenstown from 1891 to 1928 when the *America* was sold to the Anchor Line for use in the Clyde, seeing service also at Derry, as the *Seamore* and the *Ireland* was scrapped. The *Seamore* lasted until scrapped in 1945. They had 98 horse power side lever engines. The ultimate exchange of the *Paladin* with the paddler *Seamore* between the Anchor Line and the Clyde Shipping Company is indicative of the declining needs of the liner company in the 1930s.

Red Funnel at Southampton also owned the triple expansion engined *Calshot*, built in 1930. The *Calshot* was certified to carry 566 passengers in two classes centred on separate saloon accommodation. Strangely she had one oil-fired boiler and one coal-fired.

All of these vessels at some time in their career were involved in occasional excursion services, and the Red Funnel vessels were also called upon to provide additional services on the Cowes ferry. In the 1930s one shilling (three pence for children) would buy a two hour cruise from Princes Landing Stage to the Crosby Lightship in the Mersey estuary aboard the *Egerton* or the *Flying Breeze*. An annual event was the Christmas trip to the Mersey Lightships to deliver the Christmas dinner and presents from the Mersey Mission to Seamen. For this, the Captain, usually Thomas Physick or W A Newell, would be dressed in full Santa regalia!

There was one tug-tender in operation on the Manchester Ship Canal into the 1980s. This was the *Daniel Adamson* built in 1903 as *Ralph Brocklebank* for the Shropshire Union Canal & Railway. Along with the *W E Dorrington*, built in 1906, and the *Lord Stalbridge* of 1909, they took over the Ellesmere Port to Liverpool barge towing service which had provision for passengers on the tugs. They each had twin sets of compound machinery, and were built by the Tranmere Bay Development Company, Cammell Laird and the Dublin Dockyard Company respectively.

The trio was acquired by the Manchester Ship Canal Company in 1922 and refitted for their new owners. In 1929 the *Ralph Brocklebank* replaced the *Charles Galloway* (then 44 years old) as the canal tender and was completely refitted in 1936 and given the name *Daniel Adamson*, after the first Chairman of the Manchester Ship Canal Company. As such she was allowed to carry 100 passengers and was used to convey Board Members and other visitors on canal inspection trips as well as occasional excursions for schools. The *W E Dorrington* was sold for scrap in 1937 and the *Lord Stalbridge* was sold for use at Cardiff as a harbour tug in 1946, where she lasted as *The Rose* until scrapped in 1959. The *Daniel Adamson* remained in intermittent service until 1984 and later lay at a berth in the Ellesmere Port Boat Museum. Her hull is reportedly still in good condition, a legacy of a ship which rarely sailed in the salt water of the Mersey, preferring as she did the murky, contaminated and oily waters of the Ship Canal. The *Daniel Adamson* is currently being investigated as a candidate for preservation.

*An historic photograph of the tender **Daniel Adamson** (1903) and the tug **Stanlow** (1924) with the backdrop of the Runcorn Transporter Bridge, taken from the deck of the Mersey ferry **Egremont** on an excursion along the Ship Canal in June 1959.*

British Railways maintained the former Midland Railway tug tender *Wyvern*, which had been built in 1905, at Heysham for towing and tendering purposes until she was sold for scrap in 1960. Between the wars she typically operated on passenger duties between Fleetwood, Barrow and Heysham and coastal tows took her regularly to Preston, Glasson Dock, Douglas and Ayr. Duties included some excursion work but her main role before the First World War was to poach passengers from the Lancashire & Yorkshire and London & North Western Railway at Fleetwood and bring them to the Midland Railway steamers at Heysham. The *Wyvern* was also temporally deployed at Tilbury, Barrow and Troon and was occasionally to be seen on duty in the Fish Dock at Fleetwood. Originally owned by the Midland Railway, her name derived from the mythical wyvern featured in that company's coat of arms.

In later years, the London, Midland & Scottish Railway tugs *Devonshire* and *Ramsden* based at Barrow were licensed to carry passengers. These sisters were built in 1934 by Cammell Laird and were named after docks in the Barrow dock system. The *Devonshire* became *Furness* in 1937 at the request of the Bibby Line who wanted the name for a new troopship. Passenger duties included attendance at ship launches and local excursions. The *Furness* was broken up in 1964 and the *Ramsden* in 1970.

The Clyde Shipping Company commissioned two steam tenders in 1951, the *Flying Merlin* and the *Flying Buzzard*. They both had a prominent deck saloon linking the bridge and the funnel and were used principally to view launchings and other events on the river when not employed on towing duties. The *Flying Buzzard* became *Harecraig II* at Dundee in 1963 and was sold for use in Dublin in 1976 without change of name or registry. In 1983 she reverted to her original name as an exhibit at the Maryport Steamship Museum. The *Flying Merlin* left British waters in 1967 serving her new owners until sold for demolition in 1985.

The small harbour tug *Chipchase*, built as a twin screw steam tug in 1953 for the Blyth Harbour Commissioners, was also technically a tug/tender. She was equipped to carry passengers in the Blyth area particularly for port inspection work.

The other tug/tenders were sold for scrap one by one. The first to go was the *Paladin* in 1960, the *Flying Kestrel* in 1961 and both the *Romsey* and the *Flying Breeze* in 1962. The *Calshot* was sold in 1964 to a subsidiary of the Holland America Line to service their liners which then regularly called off Galway. Now complete with a new diesel engine, she continued her career as the *Galway Bay* on local ferry duties before returning to Southampton in the late 1980s for preservation.

The *Romsey* was replaced by a new *Flying Breeze*. The new vessel had been built in 1938 for the Anglo-Iranian Oil Company for use at Abadan and came home in 1955 when she was acquired by the BP Tanker Company to become their *BP Protector*. However, she lasted only six years in the Alexandra fleet at Southampton before being sold for further service under the Greek flag. She was in turn outlived by two diesel tug tenders in the Red Funnel fleet, the *Gatcombe* and *Calshot*.

*Alexandra Towing Comapny's last tender, the **Flying Breeze** (1938) attending the Cunard liner **Queen Mary** at Southampton on 30 July 1964.*

Some liner companies also operated their own tenders. Tenders were traditionally used also at Cherbourg (the White Star Line's *Traffic* and *Nomadic* of 1911 and Cunard's *Lotharingia* and *Alsatia* of 1923), on the Clyde (e.g. Anchor Line's *Paladin*), on the Mersey (Cunard's *Skirmisher* built in 1884, Canadian Pacific's *Bison* built in 1906 and White Star's *Magnetic* of 1891) and also at Cork and Galway in Ireland. The *Magnetic* was sold in 1932 to the Alexandra Towing Company who converted her from coal to oil fuel and renamed her *Ryde*. Her main role was to service the liners which were eking out a living as cruise ships during the Great Depression, and which frequently embarked passengers and stores in mid-river at Liverpool to facilitate quick turn rounds. The *Ryde* was used as an excursion vessel with a certificate for 914 passengers based at Llandudno in 1934 under the command of Captain A K Johns. In October that year she caught fire whilst moored on the Mersey above Tranmere. She was later sold for demolition. In addition, Liverpool Screw Towing & Lighterage had the *Stormcock* built in 1877 to service the Allan Line. In 1921 she was sold to owners in Cork and renamed *Morsecock* for use as a salvage vessel and occasional excursion steamer. She was scrapped in 1953.

*The British Transport Docks Board's harbour tug **Ramsden** (1934) was occasionally used for passengers when ships were being launched at the shipyard in Barrow. She is seen here at Heysham on 19 August 1967 with the former oil jetty in the background.*

The Plymouth tenders were operated for the benefit of the liner companies by the dock owners, latterly the Great Western Railway, and from 1948 onwards by British Railways. The first screw steam tender was the ***Smeaton*** which had compound machinery and was built by Laird at Birkenhead in 1883. The next was the ***Sir Richard Grenville*** built by Laird in 1891, also with twin sets of compound machinery. Gourlay Brothers of Dundee built the ***Atalanta***[1] for the London and South Western Railway in 1907 for use at Southampton. She was found to be unsuitable and was sold in 1910 to the Great Western Railway and moved to Fishguard with spells also at Plymouth. Cammell Laird built the ***Sir Francis Drake*** and ***Sir Walter Raleigh*** in 1908, effectively replacing paddle tenders of the same names. This pair comprised coal-fired twin triple expansion engined ships capable of 11 knots and were distinguishable by their very tall and raked single funnel and almost full length Bridge Deck. They maintained the tender service at Plymouth for many years, and could carry up to 1200 deck passengers.

Liner calls at Fishguard did not develop as expected and the ***Atalanta*** was sold to the Royal Mail Steam Packet Company in 1923. Again used at Southampton, her specific task was to service the new Hamburg to New York liner calls at Spithead. However, these ceased in 1926 and the ***Atalanta*** was again sold this time for use in Cherbourg as the ***La Brétonnière*** under contract to the Royal Mail Steam Packet Company. She was scuttled in 1940.

At the peak of liner calls at Plymouth (there were nearly 800 in 1930) the fleet of tenders was joined in 1929 and 1931 respectively by another pair, the ***Sir John Hawkins*** and ***Sir Richard Grenville***. These were built by Earle's Company at Hull, the former with coal-fired boilers and a slightly taller funnel, the latter oil-fired. They were licensed to carry 800 passengers on tender duties and near shore coastal excursions. The new ships displaced the ***Smeaton*** and the original ***Sir Richard Grenville***, the former going to Belfast owners until scrapped in 1947, the latter renamed ***Penlee*** to release her name for the new tender, was sold to the Dover Harbour Board and renamed ***Lady Savile***. In 1946 she was resold for static use at Leigh-on-Sea as the Essex Yacht Club's headquarters, where she remained a feature of the sea front until sold for demolition at Queenborough in 1976.

[1] Not to be confused with the turbine steamer ***Atalanta*** built in 1906 for the Glasgow & South Western Railway for use on the Clyde, and subsequently employed at Blackpool as an excursion steamer in the late 1930s.

*The **Lady Savile** (1891) in her final resting place at Leigh-on-Sea as the Essex Yacht Club's headquarters.*

The ***Sir Walter Raleigh*** was sold after World War Two. Her stern had been cut away to facilitate experimental work with torpedoes and she was deemed not worthy of repair. She was rebuilt for use at Cherbourg, however, as the ***Ingénieur Reibell*** where she replaced a tender of the same name, formerly the White Star Line's ***Traffic***, and lasted in service well into the 1960s. Her sister was withdrawn and scrapped in 1954.

The younger pair, the ***Sir John Hawkins*** and ***Sir Richard Grenville***, continued to service a declining number of liner calls interspersed by trips round the Eddystone Lighthouse. After the French Line stopped using Plymouth, the ***Sir John Hawkins*** was withdrawn in 1961 and sold for scrap the following year. Her sister lasted until October 1963 when the tender service was finally withdrawn. She was bought initially for excursion work at Torquay, but quickly resold to Jersey Lines Limited. Extensively modified and now with a passenger certificate for 550 and room for up to 15 cars, she took up duties on a Jersey to St Malo and Guernsey roster as the ***La Duchesse de Normandie***. She was affectionately known on the islands as 'Sooty', a name which reflected her ability to produce smoke at the slightest excuse! Six years later her owners went bankrupt and the old ship was dispatched to the breakers, the last of a long line of Plymouth tenders, and the last sea going oil-fired passenger steamer to operate in British waters.

La Duchesse de Normandie *(1931) in dry dock at Southampton with propeller trouble.*

Red Funnel's *Calshot*

The *Calshot* was launched at Woolston on 4 November 1929 for the Southampton, Isle of Wight & South of England Royal Mail Steam Packet Company as their largest ever tug/tender. Passenger accommodation was divided between first and second class, the first class saloon was handsomely panelled in hard wood, and furnished as befitted a first class liner passenger in the 1930s, whereas the second class saloon was a bit more spartan but nevertheless comfortably appointed. Apparently, however, for all her glamorous fittings, toilet facilities were under considerable pressure whenever a full complement of passengers was on board. Her after deck not only contained all the normal towing gear expected of a tug, but was spacious enough to carry mail bags, luggage and even occasional passenger's cars.

Passengers could view the engines from passageways on either side of the engine casing. These were only protected by partial glazing after the war, but there are no reports of passengers falling to the engine room floor! The peculiar arrangement of one coal-fired and one oil-fired boiler came to a head in 1955 when Thorneycroft were asked to quote for full conversion to oil fuel. With the quote coming in at nearly £12 000 her owners considered this an unwise investment in an old ship and she was left still with the assurance that she could survive both a coal strike and an oil shortage.

War service had initially taken her to Scapa Flow on Admiralty service, then in 1942 to the Clyde on civilian duties servicing troopships anchored at the Tail of the Bank. In 1944 the *Calshot* returned to Southampton to prepare for the D-Day landings. Returning to civilian duties for her owners at Southampton in June 1946, the *Calshot* continued to operate both as a powerful steam tug and as a passenger tender. She was given a domed cowl top to her funnel at this time. Occasionally she would also deputise on the Southampton to Ryde ferry, but as time went on these demands reduced.

The *Calshot* was replaced in 1964 by a new diesel tug/tender. She was sold, going to a subsidiary of the Holland America Line for tendering and excursion duties at Galway. Her triple expansion engines were ripped out by her new owners and replaced by diesels, her tall funnel was cut down to more modest proportions, the second class lounge was converted to crew cabins, and she was given the new name of ***Galway Bay***. Later she was bought by Galway Ferries for the Galway - Aran Islands service and on becoming redundant was bought in 1986 by Southampton City Council for conservation at Southampton under her original name and livery. She currently lies at the seaward end of the former Ocean Dock at Southampton.

*The **Calshot** is seen at Southampton in spring 1987 still bearing the name **Galway Bay**.*

(Mick Lindsay collection)

Chapter 9 : Dirty British coaster

The last of the steam coasters were withdrawn in 1983. But in 1955 the steam coaster was very much in evidence. That year eight new steam colliers were built, mainly for the Central Electricity Generating Board and the North Thames Gas Board, and one coastal tanker. The last of John Kelly's steam colliers, the ***Ballylagan***, the last of a class of four, was also delivered that year. Other types of steam coaster had all but ceased to be commissioned in the early 1950s once enough marine diesels were being produced after the war to satisfy demand, although the majority of coasters then in use were still steamers. One further coastal tanker, the ***Esso Preston***, was commissioned in 1956, William France Fenwick received the ***Helmwood*** and Stephenson Clarke the ***Arundel*** also in 1956, and Glen & Company received their last steamer the ***Winga*** in 1957. There were no more coastal ships with triple expansion machinery built for British owners although a few tank cleaning craft and other specialist harbour and inshore vessels were later commissioned. The ***Esso Preston*** was only designed as a steamship as she needed the auxiliary steam to handle cargoes of bitumen, otherwise she too would have been diesel driven.

*The **Joseph Swan** (1938) was a typical pre-war steam collier and was built by S P Austin & Son for the London Power Company. She was sunk by a German E-boat in September 1940 with the loss of 17 out of her 18 crew members.*

(Author's collection)

By 1965 the majority of the British coaster fleet was diesel engined. There was just 96 steam coasters left, mainly comprising colliers and tankers, although there were still a few traditional dry cargo ships (The Appendix lists all vessels driven by steam reciprocating engines, other than paddle steamers, registered in mid-1967). There was just one steamer left under the Irish flag, the products tanker ***Irish Holly***, which had been built for Irish Shipping Limited in 1954 and was finally withdrawn in 1967. Irish Shipping was a relative newcomer on the scene, having only been formed in 1941 by the Irish Government to ensure that Ireland, which was a neutral state in the war, could supply itself with vital imports without relying on the UK.

*The last Irish steam coaster was the **Irish Holly**, seen here at Barton on the Manchester Ship Canal in July 1966.*

By 1975 the situation was virtually one of extinction. The Central Electricity Generating Board had a diminishing fleet comprising six steam colliers built between 1950 and 1955 and four diesel driven up-river flatirons. These were managed for them by Stephenson Clarke. However, by 1977 the *Charles H Merz* and *Sir Johnstone Wright*, both built in 1955, had been sold out of the fleet for further trading under the Panamanian flag. In 1980 the *Sir John Snell* was deemed unworthy of repairs for her 25 year survey and was disposed of. Other steam colliers were sold shortly afterwards. The collier fleets gradually reduced in size as power stations and coking plants were rationalised, and the power stations were increasingly being supplied by rail or developed to burn oil or natural gas. The last steamers were the *Cliff Quay*, *James Rowan* and *Sir William Walker* all of which had received extensive repairs to tank tops just prior to their 25 year surveys in order to rectify grab damage.

*The **Charles H Merz**, outward bound in the Thames to load another cargo of coal.*

(Stuart Emery collection)

The fleet of John Kelly colliers and bulk cargo carriers stood at 26 in 1955 with only one (second-hand) motor ship. The last single hatch steamers were withdrawn from the fleet that year and the next, reflecting the hold that road transport had taken. The most remarkable withdrawal for demolition was the *Ballybeg*. She had been built by Ailsa Shipbuilding at Troon in 1898, and at 445 tons gross was a substantial ship for her day. She was in fact only the second steamer to be built for John Kelly, and fittingly was demolished near her birth place at Troon. The last of the company's steam colliers were the *Ballymena*, *Ballymoney* and *Ballylumford*, which were sold in 1971 to Shipbreaking Industries at Faslane for a mere £30 000, and the *Ballyhill* in 1973. The youngest of the steamers, the *Ballylagan*, had been lengthened, along with some of her sisters in 1956, and was sold for further trading in 1970. The last coal burner in the fleet had been the *Ballygarvey*, built as the *Donaghadee* in 1937.

The *Esso Preston* was still in operation in 1975. The Shipping & Coal Company maintained the *Highland* dating from 1951 and formerly the *Captain J M Donaldson* in the Central Electricity Generating Board fleet. Stephenson Clarke had sold the *Arundel* but acquired the small steam sludge tanker *Megstone* in 1972, albeit laid up awaiting sale in 1975. There were also three estuarine steam sludge carriers still in use, the veteran *Salford City* which had been working down the Manchester Ship Canal to dumping grounds in Liverpool Bay since 1928, the *Edward Cruse*, built in 1954 for the Greater London Council, and the *Shieldhall*, built for Glasgow City Council in 1955. The diesel engine had taken over.

*The **Esso Preston** is seen at the port after which she was named on 29 August 1964.*

(Jim McFaul)

A number of companies had a policy of motorising their fleet from the 1930s onwards. F T Everard & Sons bought the Newbury Diesel Company in the 1930s specifically to supply diesel engines and parts for their fleet of single screw motor coasters, although they also operated steamers into the 1960s. The Newbury engines came in a variety of different power designs, but were kept simple as there was a shortage of trained diesel engineers then at sea. The engine designer was Mr Kent Norris and his name was reversed in giving the engines their full title of Newbury Sirron. Other operators such as Geo. Gibson & Company and Metcalf Motor Coasters, as their name suggests, shunned the steam engine altogether, whilst a number of other companies never managed the transition from steam to diesel and disappeared altogether. Steam engineers had to retrain for diesel certificates whilst many of the older officers retired from the sea altogether.

It is interesting to look at a few of the post-war coasters as examples of the types of vessel that had been developed. Most of the coastal vessels were immaculately maintained and a number of them lasted through three and four decades. The last of the engines aft, traditional single hatch coasters were the Isle of Man Steam Packet Company's **Conister**, built in 1921 by Brown's Shipbuilding & Dry Dock Co Ltd, of Hull, as the **Abington** for Cheviot Coasters Limited, and another Manx steamer, the **Ben Maye**, which was also built in 1921, this time by John Cran & Somerville Limited at Leith, as the **Tod Head**. Later named **Kyle Rhea**, she was bought in 1955 by the Ramsey Steamship Company to service deliveries of coal and other goods to the smaller Irish ports, and given the name **Ben Maye**. They were small ships, the **Conister** being 145.0 feet (44 metres) long, whereas the **Ben Maye** was only 130.2 feet (40 metres) long. The other main difference was that the **Conister** had a single triple expansion engine which drove the ship at 10 knots and the **Ben Maye** had a compound engine which could maintain a speed of 8 knots burning about seven tonnes of coal in 24 hours. They both needed a crew of seven. The two ships lasted until 1965 and when they were withdrawn and scrapped they represented the very last of their type, once so very common around our shores and particularly active in the Irish Sea.

*A remarkable but sadly undated photograph of the Belfast-registered **Craigavad** (1924), complete with deck cargo of cars that would have been popular at the time. The era would seem to have been the mid-1960s.*

(Bernard McCall collection)

The larger type of traditional coaster, the twin hatch, amidships bridge and engines aft configuration survived a little bit longer. Typical of these were the *Craigantlet*, *Craigavad* and *Craigolive* belonging to Hugh Craig & Company which ended their days on the Preston to Belfast break bulk service until withdrawn in 1965 or the *Glenageary* and *Glencullen* which maintained coal imports to Dublin. They were built by the Lytham Shipbuilding & Engineering Company in 1920 for the Alliance & Dublin Consumers Gas Company. They were sold for demolition in 1963 and 1964, whereas the three Craigs carried on a little longer, the last representatives of another dying breed. Just prior to receipt of the **Glenageary** in 1920, the Chairman of the Alliance Gas Company made the following statement, referring to older fleet members which were due for replacement, and reflecting the disquiet that had arisen at his time just prior to the Partitioning of Ireland:

This company, and indirectly the consumers, have derived great benefit from having acquired the two steamers *Ardri* and *Braedale* which, during the past few years, have been profitably utilised for the conveyance of coal required for the manufacture of our gas. Both steamers have been many years afloat, and early last year, it was considered desirable to build a steamer specially suited to our needs to take the place of the *Braedale* which, as the market was favourable, we took the opportunity to dispose of. We approached the Dublin Dockyard Company with view to building the new coaster on our behalf, but as they were unable to accept our order, we were obliged to place the order in the hands of English builders. I am glad to say, however, satisfactory progress has been made with this new collier. She was successfully launched at Lytham on 21 February, and will be called the *Glenageary*. All the latest improvements in this type of coaster have been provided for, and we are hoping she will make her trial trip in May.

So pleased was the company with its new ship that Lytham Shipbuilding & Engineering Company was asked to supply the repeat order the following year, for the *Glencullen*. A third ship the *Glencree* was added in 1934, this time happily Vickers Ireland Dublin Dockyard was able to accept the company's order! The ships ploughed repeatedly across the Irish Sea between Dublin and Liverpool in a sedate voyage that took a minimum of eighteen hours, there being precious little indigenous coal to be had in the island of Ireland to maintain the Dublin coking plant. During the war they were occasionally strafed by machine gun fire and bombs from German aircraft fell close by – there were injuries but happily no serious losses on these neutral ships.

Life aboard these sort of ships was described by R Stead who was an AB aboard the 1923 built steamer *Slemish* in 1946, in an article which first appeared in *Sea Breezes* in 1993. On first joining the 1834 tons gross vessel at Barry and coming from deep sea vessels previously, he reported:

The fo'c'sle was divided for two watches who shared a flushed toilet which was reached from the open deck, but washing oneself, or any other social nicety like shaving had to be done using a bucket. On that first evening, in the warm summer air, we stripped off completely and washed ourselves as we stood on the foredeck. We thought it would be a kind of protest but no-one seemed to notice us amidships and the only reaction we got was from other members of the watch who made ribald comments as we stood splashing ourselves.

Even this frolic was almost spoiled by our being rationed for hot water obtained from the galley but we got more by bleeding the cylinders of the foredeck winches and catching the hot water which hissed from the valves. Later on sitting in the fo'c'sle under the oil lamps, cleaner but not a lot happier, how we complained? Four on, four off watches, overtime curtailed, no electric lighting in port, no washroom, limited hot water and having to supply and cook our own food.

Mr Stead stayed with the ship a further six months. The *Slemish* continued to operate for her owners the Shamrock Shipping Company of Larne, until sold to Thomas Leitch of London in 1956. Involved in a collision off Gravesend with Stephenson Clarke's new steam collier *Borde* that same year, she sank and was declared a constructive total loss. Her crew of 22 were safely taken off. The *Slemish* had been built by Robert Thompson & Sons of Sunderland as the *Gwentland* for Mordey, Jones & Company of Newport and had been sold to Cardiff owners in 1936 to become the *Bramhill*.

Some steam coasters had suites available for use by their owners. The oil-fired triple expansion engined steamer *Guinness*, for example, was built in 1931 to maintain the vital link between the St James's Gate Brewery in Dublin and the consumers in England. Her English terminal was London until the Park Royal Brewery opened in 1938 when she switched to Pomona Docks at Manchester. Although not fitted out for passengers she did have an owner's suite on the Main Deck forward, which was occasionally used by the Guinness family and their friends. The service continued throughout the war; each crew member receiving a pint of stout each day to discourage pilfering. The *Guinness* was withdrawn and scrapped in 1963, being replaced by a motor ship.

Porter was first exported from Dublin to England in the late eighteenth century and the brewers first became shipowners in 1913 buying the collier *W M Barkley* from John Kelly. Sunk in the Great War, the *W M Barkley* was succeeded by three more steam coasters which were ordered as self-trimming colliers for the John Kelly fleet. Two of them were bought off the stocks and the third was purchased a year later following only one years service in the coal trade. Each was converted for carrying beer casks with the appropriate cooling plant installed in their holds. These three identical steamers were the last half of an order for six steamers placed with Scott & Sons of Bowling for delivery in 1913 and 1914. One of the colliers, the *Clareisland*, was sold when the *Guinness* was commissioned, but the other two, the *Carrowdore* and *Clarecastle*, ran alongside the purpose-built *Guinness* throughout the 1930s and 1940s only being displaced in 1952 with the arrival of two new motor ships. The last of the Guinness boats was a diesel driven bulk tanker, the *Miranda Guinness*, but shipowning ceased in 1993 in the face of competition from vehicle ferries.

*The **Guinness** (1931) passes through the Barton Bridges on passage up the Manchester Ship Canal on 3 June 1962.*

With regard to the coastal tanker fleets, many of the pre-war steam tankers had been disposed of in the immediate post-war years as new war-built coastal steam tonnage became available. Some lasted longer and the former *Atheltarn*, one of three sisters built on the Mersey in 1929 for the British Molasses Company, survived as Everard's *Acclivity* between 1952 and 1966 as a specialist coastal products tanker. Last of the general steam tankers to remain in service were the Shell Mex & BP Limited *Poilo*-class dating from 1921. Five small ships in this class remained in service until the late 1960s.

The bulk of the surviving steamers were of the war time standard types, other than the post-war builds, which had principally been for the collier fleets. The standard dry cargo coasters were the *Tudor Queen*-type, and the B and C Shelter Deck Type, others having diesel engines. None of the steamers survived beyond the early 1960s under the Red Ensign, one of the last being Holderness Steamship Company's *Holdernith* which was withdrawn only in 1963. The *Icemaid*-type wartime colliers were based on the Gas, Light & Coke Company's prototype *Icemaid* built in 1936. The last to leave the UK register was the *Glanowen*, formerly the *Empire Peggotty* and bought by Harries Bros. of Swansea in 1946 and sold to W Dickinson & Company in 1964. She was used in the East Coast 'coal south and cement north' trade until sold for further service under the Greek flag in 1965.

The wartime tanker classes fared a little better with the counter stern *Empire Cadet*-class becoming a familiar sight in the 1950s and throughout much of the 1960s. Several of these ships were operated by F T Everard & Sons, others by BP and Esso as well as the Bulk Oil Steamship Company whose *Pass of Balmaha*, dating from 1933, had inspired the *Empire Cadet*-class. Quite a few of the larger TES type, designed for service in the Far East, came back to join the fleets of Shell and BP, but again none survived into the 1970s. The last of the Empire ships to be withdrawn were the Everard tankers *Alchymist*, ex *Empire Orkney*, *Argosity*, ex *Empire Lass*, *Aureity*, ex *Empire Cadet*, and motor vessel *Averity*, ex *Chant 53*. Everard's also maintained their C class estuarial tankers, all ex-wartime Ministry of Defence oilers, until the early 1970s.

An interesting quartet of coastal tankers were the shallow draught trunk-decked steamers *Esso Chelsea*, *Esso Fulham*, *Esso Lambeth* and *Esso Wandsworth*. They had been built in 1945 by the Bethlehem Shipbuilding Company, Maryland, and the Barnes-Duluth Shipbuilding Company, Minnesota as part of ten identical American war-builds designed to maintain the vital export of oil from Lake Maracaibo in Venezuela to the oil refinery at Aruba in the West Indies. The Esso ships were built as the *Amacuro*, *Trujillo*, *Caripito* and *Guarico* respectively. The ships were originally part of the Lago Fleet and were managed by Andrew Weir & Company with British crews, but owned by the Standard Oil Company of New Jersey. In 1948 a pipeline was established between the lake and a new terminal on the Paraguana Peninsula and the steam tankers were no longer required to bounce fully laden over the bar into the Atlantic rollers.

*The former Lake Maracaibo steam tanker **Esso Chelsea** (1945) seen at Barton on the Manchester Ship Canal in March 1965.*

The four ships that transferred to the UK coastal fleet of Esso Petroleum in 1956 promptly developed a reputation for poor handling in confined waters despite twin triple expansion engines and the fact they were designed for confined and shallow waters in the first place! The quartet were affectionately referred to by Esso crews as the 'Dingbats'. They were sold for scrap between 1965 and 1970, the *Esso Wandsworth* following a collision in the Thames in 1965. The last to be scrapped was the *Esso Fulham*.

The very last steam coaster to be built, and coal-fired at that, was the 160 gross ton Clyde puffer *Stormlight*. She was built at Northwich by W J Yarwood in 1957 for Ross & Marshall which merged later that year with Hay Hamilton to form Glenlight Shipping. The *Stormlight* was fitted with a two-cylinder compound expansion engine, could carry 160 tons of cargo and used 4 tons of coal per steaming day. She was withdrawn from commercial service in the early 1970s.

Still in operation at that time was the veteran *Mellite*, the very last of the old-style puffers and little more than a seagoing barge, which was used in her later career to carry water out to ships at the Tail of the Bank from Greenock. The *Mellite* was always immaculately kept with lots of polished brass about the boiler casing. She was built as a dumb barge in Port Glasgow in 1873 and converted to a steamer late in the nineteenth century with a second-hand compound engine taken from a steam yacht. Unlike most puffers she had a return tube Scotch boiler rather than a steam crane vertical type.

There are many more steam coasters that deserve mention either in the last of the steamers context or as special or distinctive vessels. Their story is told in other books about coastal steamers so that these vessels and their crews will not be forgotten.

Stephenson Clarke Shipping Limited

Stephenson Clarke & Company, affectionately known to seafarers as Stevie Clarke, was formed in 1850, although the company's roots go back into the early eighteenth century when Ralph and Robert Clarke purchased shares in a 300 ton sailing vessel. The core business of Stephenson Clarke was the transport of coal from North East England to the Thames. The first three steamships were ordered by the company for delivery in 1865. Fleet expansion followed until an association with the Powell Duffryn Steam Coal Company led to that company acquiring all the ordinary share capital of Stephenson Clarke in 1928. (Powell Duffryn similarly owned John Kelly Limited, whose four remaining motorships were absorbed into the Stephenson Clarke fleet in 1990). From 1928 onwards the black funnel with a broad silver band and a fleet nomenclature of southern English towns became a familiar site in the north east, Thames area and south coast ports.

So successful did the company become that it was soon put in charge of managing the main gas and power utility collier fleets. In post-war years, when these had been nationalised, they comprised the Central Electricity Generating Board, the North Thames Gas Board and the South Eastern Gas Board. In 1946 Stephenson Clarke jointly formed Coastwise Colliers with Wm. France Fenwick to undertake long-term charters to the County of London Electricity Supply Company. The company was dissolved after the nationalisation of the electricity supply industry in 1948 and the ships reverted to their respective former owners.

The first motor ship joined the fleet in 1947 and the first tankers in 1957. By 1956 Stephenson Clarke operated 16 steamers and 14 motor ships, the steamers ranging from 3 to 27 years in age. During the 1960s, all the coking plants were shut down as North Sea gas came on stream. This, coupled with the closure of the smaller power generating plants in favour of big units with deep sea jetties, made many of the smaller colliers redundant and put an end to the less economic steam engined colliers. From then on Stephenson Clarke diversified away from the coal trade. Powell Duffryn sold its last shares in the company in 1994, but Stephenson Clarke remains involved to this day with the traditional small-port bulk cargo business.

The last steamers in the fleet were the **Arundel**, **Borde**, **Bramber** and **Heyshott**. They were built between 1953 and 1956. Only the **Arundel** survived into the early 1970s, leaving the post-war built steam colliers in the Central Electricity Generating Board fleet as the final vestige of this kind of ship. However, they too were soon withdrawn, and by mid-1977 none was left in the UK registry.

The **Bramber** outward bound in the River Thames.

(Stuart Emery collection)

Chapter 10 : Estuarine steamers

Although the paddle steamer is suited for use at relatively shallow landings, the screw steamer was best suited for cross-river purposes. Wherever there was a running tide, the screw steamer could nose into its riverside berth without difficulty, only having to use its engines to reverse and position itself at times of slack water. Most of these ferries were small such as the Tilbury-Gravesend passenger ferries *Rose*, *Catherine*, *Edith* and *Gertrude* (see also Chapter 8). Others were built as vehicle ferries and were substantially larger, e.g. the Mersey luggage boats or the Shields ferries *Tynesider* and *Northumbrian*. There were, however, a large number of them, diminishing in the 1950s and early 1960s as they were either withdrawn in response to new bridges and tunnels, or converted to diesel.

Perhaps one of the best known fleets of estuarine ferries is that of the Mersey ferries. Traditionally these were divided between the passenger ferries and the luggage boats, and were operated by a variety of owners, latterly the Birkenhead (Woodside) ferries of Birkenhead Corporation and the Wallasey Ferries (Seacombe and formerly also Egremont and New Brighton) to Liverpool (Princes Landing Stage). Despite the Mersey Railway tunnel which was opened in 1886, and the Mersey Road tunnel which dates from 1934, the ferries, albeit diesel ones, remain in demand to this day. The coming of the road tunnel did, however, eventually put paid to the luggage boats. Long queues of traffic used to develop during the day for the ferries prior to the tunnel, whereas after its opening only horse-drawn vehicles and petrol carriers needed the luggage boats until this service was finally withdrawn in 1947. All of the luggage boats were coal fired.

The last of the Birkenhead luggage boats were the *Barnston*, *Churton*, *Bebington* and *Oxton* built between 1921 and 1925, and the *Prenton* built in 1906. They followed the design of earlier boats with twin fore and aft propellers driven by a pair of four-cylinder triple expansion engines. Three ships were generally in steam at any one time. On the opening of the Road Tunnel, the *Prenton* was sold for scrap and the others soldiered on until the Birkenhead service was withdrawn in 1939. Two of the remaining ships then went to Dutch owners and the other two were converted in the war into crane barges for unloading aircraft from cargo ships.

*Seen from an angle which emphasises the utilitarian looks of the luggage boats is Birkenhead Corporation's **Bebington** of 1925.*

(Author's collection)

On the Seacombe service the elderly Victorian *Wallasey* and Edwardian *Seacombe* were replaced by the new luggage boats *Liscard* and *Leasowe* in 1921. They were characterised by large white and black topped funnels overlooking clear and unobstructed vehicle decks. This pair remained at Liverpool during the war but were sold shortly afterwards. The very last of the luggage boats was the *Perch Rock*, delivered from the Caledon Shipyard at Dundee in 1929. She was also the largest with gross tonnage of 766. The *Perch Rock* had the distinction of closing the vehicle service in March 1947.

*The Wallasey luggage boat **Perch Rock** (1929).*

(Keith P Lewis)

She was sold in 1953 following limited use as a passenger ferry to Swedish owners and converted for use as a train ferry conveying sugar beet trucks. After only one year as a train ferry she was used as a vehicle and passenger ferry, largely rebuilt in 1960, she maintained the Helsingborg to Helsingor ferry until 1970.

The last of the Birkenhead passenger steam ferries were the quintet *Hinderton*, *Upton*, *Thurstaston*, *Claughton* and *Bidston* which were delivered between 1925 and 1933 by Cammell Laird & Company. They were distinctive in that the Promenade Deck was extended to the full width of the ship and they had the classic three cab type bridge structure. Steam propulsion was preferred for these ships because of the great number of engine manoeuvres that are needed on ferry duties, as diesels would have required large capacity compressed air start reservoirs and could be dangerous if their engines failed to fire during a reversing manoeuvre. The steamers had twin screws and twin rudders and were coal fired. The *Upton* was sold for service at Southampton in 1946 (Chapter 8), whilst the other four became the mainstay of the post-war service until they were displaced by a new diesel trio in the early 1960s, the *Claughton* being the last steamer to remain in service and only going for scrap in 1962. Three diesel vessels now maintain the service, each having recently been extensively refurbished, re-engined and partly rebuilt for modern day use.

*The simple layout of the Birkenhead ferry **Thurstaston** (1930) included three separate bridge cabs by the funnel.*

(Author's collection)

Between the wars several large capacity ferries were also commissioned for the Wallasey services, namely the *Francis Storey* and *J Farley* in 1922, the *Wallasey* and *Marlowe* in 1927 and the *Royal Iris II* and *Royal Daffodil II* in 1932 and 1934 respectively. The suffix II was needed as the old *Royal Daffodil* had been sold the previous year to the New Medway Steam Packet Company without conditions on using the royal name - a name given to the *Daffodil* and *Iris* by King George V in recognition of the legendary exploits of these two vessels at Zeebrugge in 1918, so allowing them to adopt the names *Royal Daffodil* and *Royal Iris* thereafter.

The *J Farley* and *Francis Storey* were commissioned with four inch mourning bands painted around their bulwarks, their namesakes having both died shortly beforehand. They maintained the New Brighton service in the 1930s. Used by the Admiralty in the war to handle torpedo nets, they returned to the Mersey to be sold in the early 1950s. The *J Farley* was converted to burn oil in 1948, for use at Portland by the Admiralty, where she remained until the mid-1980s, and the *Francis Storey*, renamed *Killarney*, was used as a tender at Cork, being scrapped only in 1960.

The *Wallasey* and *Marlowe* were big passenger carriers, certified for 2233 passengers on ferry duties. They had twin Flettner rudders, as well as twin screws and an innovation was the cruiser stern. The *Royal Iris II* and the *Royal Daffodil II* could carry even more passengers as they had a third deck aft of the funnel instead of the awning carried by the earlier vessels. The *Royal Iris II* was built with cruising in mind and had Tudor-style hard wood panelled main lounges and a dance floor. The *Royal Iris II* became *Royal Iris* in 1947 when her predecessor was renamed *Blarney* (as a Cork tender) but the *Royal Iris* was again renamed in 1950 as the *St Hilary* in anticipation of the commissioning of the new diesel-electric *Royal Iris*. The *St Hilary* later became the Dutch ferry *Haringvliet*, and was scrapped in 1956.

Despite the *Royal Daffodil II* spending 13 months of the war on the bottom of the Mersey alongside the Seacombe Stage following bomb damage, she survived, having lost only her teak panelling in the lounge. She reopened the river cruise services painted white overall with a two tone yellow funnel in 1947, but reverted to black hull when the new diesel-electric cruise boat *Royal Iris* arrived in 1950. In 1957 the *Royal Daffodil II*, like her sister before her, was also renamed *St Hilary* and later scrapped in 1962, the previous *St Hilary*, formerly the *Royal Iris*, having in the meantime been sold.

The distinction of the last steam ferry on the Mersey fell to the *Wallasey*, which was withdrawn and sold for demolition only in 1964. Her bell is preserved in Wallasey Town Hall. The *Marlowe* was scrapped in 1958. However, the changeover from coal-fired steamer to diesel was not undertaken lightly by Birkenhead Corporation. Despite Wallasey adopting diesel for all replacement tonnage after the war, the Birkenhead ferries manager announced in 1951 to the Tunnel Committee as reported by T B Maund in his treatise on Mersey Ferries (1991):

That with steam there had been one breakdown in service in 25 years, whilst diesel engines were considered to be unreliable, thereby making it essential to install more than one engine. Reference was made to two new pilot boats, each with two engines, and the *Royal Iris* with four.

A motor vessel, would, however, require a crew of only six compared with the current level of nine but there were other problems. Since the opening of the tunnel and the closing of the Rock Ferry and goods services all the younger members of staff have obtained work elsewhere, leaving only the senior officers and casual labour. All have been in our service since boys and none hold Ministry of Transport Certificates, hence three engineers would have to be engaged for each proposed motor vessel and they would demand higher salaries for greater qualifications. The workshop and its staff were unsuitable for diesel engine maintenance and this work would need to be contracted out. The manager concluded with an observation that a dozen new tugs were all steam propelled 'a fact that gives rise to considerable speculation regarding the advisability of installing diesel engines in Mersey river craft'.

In 1958 the Ferries Committee finally abandoned their make-do-and-mend policy when the old ferry manager Mr Cowan was succeeded by the bus manager George Cherry, and an order was placed for the first two Birkenhead diesel ferries. These were duly delivered by Philip & Son at Dartmouth for the sum of £482 278.

On the Tyne, the North Shields to South Shields ferry stayed in steam throughout the 1960s. The twin screw vehicle and passenger ferry *Northumbrian* was the last steamer, and for many years she had been partnered by her older near-sister the *Tynesider*. The *Tynesider* was a product of Philip & Sons at Dartmouth and was delivered to the Tyne in 1925. The *Northumbrian* came to the Tyne Improvement Commission from Hawthorne Leslie in 1930. They were larger versions of the Jarrow ferry *A B Gowan*, with oversize funnels and rounded hull shapes that maximised payload for minimum draft. They had a passenger lounge forward of the Car Deck and beneath the open Promenade Deck which contained the bridge cab and a single mast. In their later years they passed into the ownership of the Tyne Port Authority. The *Northumbrian* finally retired with the arrival of a new diesel ferry in 1976.

There were many smaller steam reciprocating engined passenger ferries which operated a variety of cross river services. The last of the Portsmouth to Gosport ferries to remain in steam was the *Venus* which had been built in 1948 and had been equipped with a compound engine. Sold to Barham & Hogg in 1967 she was then managed as part of Favourite Boat Cruises at Southampton, being converted to diesel propulsion only in 1969 ready for the 1970 season. Although the vessel ended her days as a sailing cruiser in Australian waters, her original engine is on show in the Southampton Maritime Museum. The last steamer built for the Portsmouth Harbour Ferry Company was the *Ferry Princess* in 1959, but converted to diesel in due course by her owners. A number of former Gosport steam ferries were converted to diesel for use as excursion vessels notably on the upper Thames, and at least one, the *Vesta*, is still in service.

*One of the last steamers in the Portsmouth Ferry Company's fleet was the **Venus** (1947) seen at Southampton as an excursion steamer before conversion to diesel in the winter of 1966.*

The Fleetwood to Knott End steam ferry across the Wyre estuary came to a tragic end in 1957 after a boiler explosion and the death of three men. The service had been maintained by the *Wyresdale* since her delivery in 1925 by local shipbuilder James Robertson & Sons (Fleetwood). The little *Wyresdale* had twin compound engines and twin screws and had coal-fired boilers. She was distinguished in that she carried hinged gangways amidships to facilitate rapid embarkation and disembarkation. Since her demise in 1957, a variety of motor ferries has operated the service, although in recent years the service has been intermittent.

On the Thames, the twin screw coal-fired *Tessa* and *Mimie* maintained the Tilbury to Gravesend vehicle ferry until it was withdrawn in 1961. They were built in 1924 and 1927 respectively for the London, Midland & Scottish Railway and each had a pair of twin cylinder compound engines. They could accommodate up to 30 cars and their passengers, the latter obliged to stay with their vehicles for the short crossing. The memorable feature of the crossing was always the banter with the ticket seller as he came round the vehicle deck and the total silence from the ship and her engines as she once again pulled effortlessly into the tideway.

The Tilbury-Gravesend passenger ferries were the *Rose*, *Edith*, *Catherine* and *Gertrude*, built by Robertson & Company in London between 1901 and 1911, and developed on the model of the *Carlotta* of 1893, but which was withdrawn and scrapped in 1930. These vessels were also equipped with twin two-cylinder compound engines and were each of just greater than 250 tons gross. Their original owners were the London, Tilbury and Southend Railway which was taken over in 1912 by the Midland Railway, and in 1923 by the London, Midland & Scottish Railway. The *Gertrude* was sold to the New Medway Steam Packet Company as the *Rochester Queen* in 1932 (Chapter 8), but the other three soldiered on until replaced by a trio of diesels in 1960/1961.

The Clyde Navigation Trust operated a variety of ferries on the river. Three distinctive steamers were the *Vehicular Ferry-Boat No 1*, *Vehicular Ferry-Boat No 2* and *Vehicular Ferry-Boat No 3*; the fourth was diesel-electric. Built for use at Finnieston and Govan, they had a steam powered hoistable platform to suit the tide, on which passengers and vehicles were carried. Built in 1890, 1900 and 1908 respectively, and equipped with triple expansion engines and propellers fore and aft, these ships served until the 1960s when they were withdrawn and scrapped.

A variety of steam bollard to wire pulled ferries have operated such as the *King Harry Ferry* on the Fal and the former Erskine ferries *Erskine* and *Renfrew* dating from 1925 which pulled themselves back and forth on a fixed chain. The King Harry Ferry was replaced by a diesel-electric unit in 1956, and the Erskine Road Bridge opened in 1971. Probably the earliest replacement of a fixed wire steam ferry on the opening of a bridge was the Walney Island ferry at Barrow in Furness, which ran from 1878, with a new and larger vessel installed in 1902, but was closed in 1908 when a bridge was completed.

Chapter 11 : Death by diesel

The very fast triple expansion engined ferries of the late Victorian and Edwardian era were never repeated because of the arrival of the more efficient steam turbine engine. As early as 1904 the direct-drive steam turbine steamer was showing its paces in the Dover Strait and on the Stranraer to Larne route. The intermediate type steam ferries, however, survived this onslaught only to start facing competition from the diesel engine from the 1930s onwards.

The diesel had a very slow passage into acceptance despite it being smaller and lighter in weight than the conventional triple expansion engine. There were a number of reasons why its acceptance was slow, not least conservatism on the part of the owners who were, for the most part, content with the tried and tested steam technology. A significant drawback in the early days was the lack of trained diesel engineers, and this was the reason why the very early group of motor coasters built for Paton & Hendry just before the Great War failed to survive.

The key landmark in diesel acceptance was the introduction of the *Ulster Monarch*-class of overnight passenger and cargo ferries from 1929 onwards on the Irish Sea routes. This was followed by the introduction of diesel and electric coupling for David MacBrayne's *Lochfyne* in 1931. This arrangement gave the bridge direct control of the speed and direction of the ship whenever required and overcame the need to shut down the engine and restart it in reverse whilst manoeuvring. Each engine manoeuvre required compressed air to restart the engine, and this was of finite supply in the early diesels. From the early 1930s onwards the diesel engine began to creep into the coasting trades. Initially led by the Coast Lines Group on the Irish Sea, it soon began to find favour with other operators such as the fleets of small tankers operated by Metcalfe Motor Coasters and Christopher Rowbotham & Sons.

*The world's first cross-channel diesel ferry, the **Ulster Monarch** of 1929, seen passing through Princes Half Tide Dock on arrival in Liverpool in 1966.*

*David MacBrayne's **Lochfyne** utilised the diesel with electric coupling to overcome manoeuvring and control difficulties of early marine diesels.*

*The **Cambria** was only the second main line diesel ferry to be built for the railways. She is seen off Holyhead in September 1974.*

The introduction of the Voith Schneider cycloidal propulsion system of vertically mounted blades was introduced in the late 1930s and revolutionised estuarine ferry services. Although set back by the lack of spares during and immediately after the war, the pioneer diesel Voith Schneider ferries **Vecta** at Southampton, the Southern Railway's **Lymington** also on Isle of Wight ferry duties and the Dundee to Newport ferry **Abercraig** hastened the end of the inshore steam ferry. Lack of spares from the German manufacturer of the propulsion units saw the **Vecta** laid up for much of the war and in 1946 she was converted to a conventional diesel electric screw driven ferry. Notwithstanding, the huge advantages of manoeuvrability with the Voith Schneider system provided a considerable fillip to the adoption of diesel power for these craft and there were no steam ferries of note built after the war.

The popularity of fuel oil over coal had increased in 1926 during the miners' strike and from then on coal was less favoured by all but the collier owners. It was for this reason that the Southampton tender **Calshot** was commissioned with one coal-fired and one oil-fired boiler, hedging bets against a shortage of either commodity. In later years fuel oil, of course, was widely used and only became a problem for the steamers after the Arab-Israeli conflict in the winter of 1973/74 when the price of heavy fuel oil increased fourfold in the space of four months. This was the death knell for many of the oil-fired steamers, and although costs were increased also for the diesel ships, their greater efficiency limited the price hike of fuelling motor ships to more manageable levels.

The steam engine had enjoyed one small resurgence. In the austere years of the late 1940s and early 1950s, there were insufficient diesel engines and engine parts being manufactured to satisfy demands from the shipowners who were anxious to bring their fleets up to pre-war strengths. The resurgence might have been considerably larger but for the hike in prices between 1939 to 1946 – the cost of building a steam coaster having all but doubled simply as a result of supply versus demand. The Recession was very much a thing of the past.

During the war itself, much of the Empire-type building of coastal ships, mainly tankers and colliers, was diesel driven. Many ships designed for diesel were of necessity equipped with triple expansion engines whenever diesels were in short supply. That being so there was an increase in the provision of diesel engined coastal ships as the war progressed, and although the Ministry of War Transport then owned all of these ships, they were put under the management of commercial operators. Many of these operators had not previously operated motor ships, but almost all of them were impressed by the capability and efficiency of the diesel – another nail was being driven into the triple expansion engine's coffin.

*A typical early diesel-driven coastal tanker was Rowbotham's **Steersman** (1936), seen entering the River Hull on 9 January 1967.*

The down-river collier fleets did not readily adopt the diesel engine. This was not surprising as the steam collier was able to bunker with long-term contracted cheap coal at each return to the coalfield ports. The vulnerability of the East Coast colliers to U Boat attack had left a greatly depleted fleet of colliers at the end of the war. Despite the threat of nationalisation of the electricity and gas companies (which occurred in 1948 and 1949 respectively), a massive rebuilding programme was needed. The preference was for coal-fired steamers for down-river colliers, and for diesel engines for the up river flatirons. In the case of the up-river vessels the smaller space required for diesel engines and bunkers optimised the payload in vessels whose dimensions, especially air draught, were otherwise constrained by the Thames bridges.

The wisdom of coal-fired boilers for the post-war down-river colliers was questioned in the 1960s when the Central Electricity Generating Board began a programme of converting its coal burners to oil fuel. Nonetheless, they remained faithful to the triple expansion engine into the 1980s, partly because of the large number of war replacement steam colliers built in the 1950s that were good for thirty years of service, and partly because the collier fleets maintained a tradition of steam engineering and expertise. It is perhaps ironic that the industry that was championing nuclear energy and gas-fired power stations should retain the steam reciprocating engine to the last; that being said, the steam turbine remains to this day the means of driving the power generators, no matter what the fuel source.

Traditional coal bunkering facilities were gradually withdrawn as coal-fired ships became more scarce. This hastened the demise of many of the coal-fired coasters employed in general tramping trades that took them away from the traditional coal loading ports.

*Typical of the post-war diesel vessels that were built to replace railway steamships was the **York** (1959) of Associated Humber Lines.*

*Successors to the "green parrots" of Ellerman's Wilson Line were diesel vessels such as the **Rapallo** (1960), a product of Henry Robb's yard at Leith.*

The diesel was introduced to the tug fleets initially through the small harbour tug, and was particularly popular with the lighter handling tug operators on the Thames. Indeed, so successful was the engine that many of the smaller steam tugs were progressively re-engined with diesels. Interestingly, Steel and Bennie converted their steam tug *Chieftain* to diesel in the late 1950s, retaining the Scotch boiler as the fuel tank and to maintain stability and trim. Larger diesel ship handling units were built for companies such as Gaselee & Knight on the Thames, the Clyde Shipping Company and the Manchester Ship Canal Company, but these were exceptional. Most post-war tug building stayed faithful to the steam reciprocating engine and it was only in the 1960s that motor tugs became the norm.

*The last steam tugs to be commissioned were the **North Buoy** and **North Wall** for Alexandra Towing in 1959. The latter was photographed as she headed out of Swansea at speed in May 1971.*

(John Wiltshire)

The last steam coaster to be commissioned was the Clyde puffer *Stormlight* in 1968 (Chapter 9). She only remained in active commercial service until 1973 when the economies of her diesel running mates and the increased cost of fuel oil pushed her into early retirement. The last steam coastal vessel to remain in service was the sludge carrier *Shieldhall* (Chapter 12), although a number of steam dredgers and tank cleaning vessels remained in service in estuarine waters for some years after that. It is interesting to note that UK operators continued to build fast steam turbine ferries for a decade after they had given up the triple expansion engine for its coaster fleets. The last turbine steamer was the Isle of Man Steam Packet Company's *Ben-my-Chree* which was commissioned as the second of a pair of steamers in 1966 and withdrawn only in 1985 – the same year that the *Shieldhall* was withdrawn.

As with the change over from steam to electricity as the power source ashore, the final transition from steam to diesel at sea, like so many historical changes, went unremarked. But the triple expansion engine was no longer the power house of our coastal passenger and cargo fleets. The concluding chapter in its demise was hastened by three factors:

the ever improving efficiency of the internal combustion engine for all sizes of vessel and speeds of operation,
the spiralling fuel costs of the mid-1970s, and finally
the onset of the roll-on/roll-off vehicle freight and passenger ferry.

The roll-on/roll-off vehicle ferry had been pioneered by Frank Bustard with the early services out of Preston and Tilbury using former Landing Ships Tank in the late 1940s (Chapter 6). Acceptance of the door-to-door road vehicle concept was slow, although some companies such as Pilkington's Glass of St Helens, realised that this was an important way to penetrate the Irish and Continental markets right from the very beginning. Lifting of UK freight vehicle restrictions on the Continent in the early 1950s helped, but it was only in the 1960s that vehicle and cargo shipment was finally recognised as the way forward. As the vehicle freight concept took hold, so the traditional break-bulk system fell into disuse and the steam coaster was finally retired for ever.

The fast passenger paddle and screw steamer had died at the hands of the Edwardian turbine steamer. The intermediate passenger steamer had died at the hands of the diesel engine. The final death roll of the traditional coasting steamer was the rise in fuel costs coupled with the arrival of the vehicle ferry. The steamer was no more.

Chapter 12 : In steam again

Today we have two active sea-going steamships, the *Shieldhall*, a former Clyde sludge carrier based at Southampton, and the perhaps better-known *Waverley*.

Manchester had the *Mancunium*, a repeat of an earlier ship of the same name built in 1933 and sunk by a mine in 1941. The new *Mancunium* was built by Ferguson Brothers at Port Glasgow in 1946, and like her earlier namesake was equipped with twin triple expansion engines, and was designed to take the city's sewage sludge to dumping grounds in the Irish Sea. Her consort for many years was the *City of Salford*. Converted to diesel in 1961, the *Mancunium* was later scrapped in 1990. Very occasionally the pair carried passengers on their run down the Ship Canal to dumping grounds in Liverpool Bay, albeit usually council officials accompanied by their families.

London had the steamer *Edward Cruse* built as late as 1954, along with a fleet of three newer diesel sludge carriers. Belfast had the *Divis*, built in 1928 for Belfast Corporation by local company Workman Clark & Company, which survived a fifty year career with her original coal-fired boilers and twin triple expansion engines, the latter supplied by her builders. And there were others serving east and south coast ports as well as the Avon.

Glasgow, of course, had the *Shieldhall*, the very last of the steam sludge carriers to be constructed. She was built by the specialist dredger company Lobnitz & Company at Renfrew specifically to carry sludge from her owners Shieldhall Works to the lower Clyde estuary. Unlike other areas where sludge carriers were active, Glasgow Corporation had magnanimously provided free saloon accommodation for selected groups of passengers to enjoy the return trip during the summer months since the conclusion of the Great War. Initially targeted at convalescing servicemen, it was later widened to include school children and the city's underprivileged, not least its pensioners. Typically the ten hour voyage would start at 0830, although without passengers in the winter this was brought forward to 0730.

The first Glasgow sludge boat was the single engined *Dalmuir*, commissioned in 1904 coincident with the opening of the Dalmuir sewage works. She was joined by a slightly larger ship, the first *Shieldhall* in 1910. The *Dalmuir* was disposed of in 1922 and ultimately replaced by a new ship, the *Dalmarnock*, in 1925. A two-ship service was maintained thereafter, save for the period 1941 to 1947 when the *Shieldhall* was loaned to the City of Manchester to stand in for their first *Mancunium* which had been mined. During this period the *Dalmarnock* maintained a double daily run, requiring round the clock manning, and only travelled as far as Loch Long in the shelter of the boom defences.

*The **Dalmarnock** (1925) seen off Gourock on 23 August 1970 as she returned from the dumping grounds in the Clyde estuary. She was withdrawn a year later.*

Eventually, in 1955 the old *Shieldhall*, then 45 years of age, was replaced by the *Shieldhall* we know today. The new ship was provided with bench seats on the foredeck and seating in the saloon and was issued with a day passenger certificate for 80. Parties had to provide their own catering but there was a large dining saloon and adjacent galley facilities, the latter often manned by caterers from schools and old people's homes. The *Dalmarnock* ceased to offer passenger facilities once the *Shieldhall* had established herself on the run. Perhaps the most remarkable thing about the two ships was their clean and spotless appearance with no escape of odour from the sludge tanks below, even whilst discharging off Garroch Head at the south end of the Isle of Bute.

In 1975 the metropolitan responsibility for water and sewerage passed to Strathclyde Regional Council. Passenger facilities were withdrawn and in October the following year the *Shieldhall* was sold to Southern Water Authority to be based at Southampton. There she was laid up for three years as her new owners had an outstanding contract for the provision of sludge disposal at sea. The *Shieldhall* was nevertheless back in steam in 1979 with a crew of twelve, this time taking sludge from Southampton sewage treatment works to dumping grounds six miles south of the Nab Tower. In July 1985 she was again laid up, somewhat suddenly, and the sewage disposal service was once again outsourced. It is presumed that her owners realised that the steamer was an expensive item to maintain. However, they also acknowledged the heritage value of their vessel, and she was placed under the management of Southampton City Museum as a candidate for preservation. The Solent Steam Packet Company was formed in 1987 and her ownership was taken from Southern Water Authority for the sum of £25,000. Proof of interest in the vessel was given in August 1985 when the museum held an open day which was attended by 1000 visitors. The attraction of a part-rivetted and part-welded hull, Scotch boilers, steam steering engine, traditional wheelhouse as well as twin triple expansion engines (15, 25 and 40 inch diameter cylinders with a 30 inch stroke) was compelling.

The *Shieldhall* was an ideal candidate for preservation because she had been maintained to a very high standard and at the age of thirty was still in excellent condition. Although some remedial work and refurbishment was necessary to maintain her in steam, the main effort has been routine maintenance over the winter periods, including soot removal from boilers, filters and exhausts, to enable the vessel to be active during the summer months, and to complete annual survey inspections and certification. Just rodding the boiler tubes is a job that takes several weeks of the winter lay-up.

*The **Shieldhall** (1955) off the Town Quay at Southampton during excursion duties on Saturday 18 July 1998.*

Getting up steam has to be done slowly and carefully in order to apply equal stress to the complex array of steam tubes in the boilers. The boilers are encased in one inch thick riveted steel plate which remains warm to the touch a week after the boilers have been shut down! The starting up procedure requires at least two inches of water to be showing in the boiler gauge glasses, which are made of hardened soda-glass. Each oil burner is carefully cleaned of soot and washed in diesel. When raising steam from cold, the initial firing is with diesel until the heavy fuel oil is warmed sufficiently to an optimum 98°C for it to thin and vaporise in the burners, and for there to be enough steam to pump the bunker fuel to the boiler and (hopefully) to start the forced draught fan.

A variety of different burners are used for the different fuel and starting temperatures. Electric forced draught has to be balanced with fuel input: too much fuel or not enough air, the exhaust gases are black, too much air they are grey. Once up and running, the response to the bridge telegraph (the port bell being a higher pitch than the starboard) is 75 psi across the high pressure cylinder for full speed ahead at 86 revolutions per minute to give the ship a speed of 9 knots (originally designed for 120 revolutions per minute to provide a speed of 13 knots). Half speed is given at 50 psi, 25 psi for slow and 12 psi for dead slow. A key part of the engine room watch is checking and monitoring valves, gauges and oiling points. Nowadays the *Shieldhall* burns one tonne of fuel per hour in steam, so that a six hour cruise can cost up to £2 500 in fuel oil alone (at 2003 prices). That the ship was economical in fuel when built reflects the price hikes in the energy sector since the 1970s. And what is the temperature in the boiler room? - Apparently 45 to 50º C is not uncommon when running at full speed for any length of time.

The *Shieldhall* is very much of traditional design. Some 82 metres long by 14 metres breadth, she draws 4 metres in ballast. She is subdivided into five watertight sections. There is a single tier of deck housing extending the full after half of the ship's length, raised forecastle, large cowled funnel and gentle sheer, the ship has a low profile and gentle gracefulness unexpected of a sewage sludge boat. The cargo space is divided into two sludge tanks divided on the centre line, and the lowest point of which is above the loaded arrival trimmed water line, i.e. full fuel and water tanks, water in boilers, stores and crew and effects but no cargo or ballast. The buoyancy spaces below are similarly divided. The sludge could, therefore, be discharged at the turn of a valve (four valves to port and four to starboard) without any need for pumping. The sludge discharged through openings in the bottom shell plating, and the valves were controlled manually from the Main Deck. On discharge, sea water was allowed into the buoyancy tanks in order to ensure adequate submergence of the propellers for the return journey.

Today the *Shieldhall* remains the last sea going triple expansion engined screw driven coastal vessel in steam. An unlikely candidate for preservation, she is nevertheless a welcome one, and her generally excellent condition is a reflection of the craftsmanship of her builders and of the men who have had the task of maintaining her and looking after her these last fifty years. This is especially true of the dedication of the volunteer labour that has maintained the ship since her transfer to the Solent Steam Packet Company. It should be noted that the practice of sewage sludge disposal to sea ceased in 1997 at the direction of the European Community.

At the other end of the country, that other stalwart the *Sir Walter Scott*, twice the age of the *Shieldhall*, continues to ply Loch Katrine and the tourist trade. She was built at a cost of £4 250 (the cost of the engines and boilers was extra). Additional transport costs to the loch meant that Denny's lost £112 on the contract they had as sub-contractors to Mathew Paul & Company of Dumbarton with whom the order for the 'passenger excursion launch' had originally been placed. The old lady has been reboilered twice (1956 and 1991) the first time with boilers built by Marshall & Anderson of Motherwell and the second by Cochran & Company of Annan. She was converted from coal to smokeless fuel in 1967. Her engines are the original set, with 8, 13 and 19 inch cylinders and a 12 inch stroke and jet condenser, and the operating pressure is 160 psi. The engines were given a complete overhaul by Waverley Excursions' staff in the winter of 1998/99.

The *Sir Walter Scott* now has a crew of six, captain, mate, engineer, stoker and two deck hands. She is licensed to carry 250 passengers on a Class V Certificate. Her passenger complement has been reduced from 416 in recent years due to Safety of Life at Sea regulations.

Elsewhere, the opportunity to travel by inland steamer is also offered on Windermere and Coniston in the Lake District. These are smaller vessels that rightly remind us of days gone by but do not quite stir the same emotions that the *Shieldhall* and the *Sir Walter Scott* are capable of doing. One observer, for example, recently described a trip on Loch Katrine as 'the quite rhythmic pulse of the steamer's engine and the wake that ripples along her fine lined hull'!

How long Sir Walter Scott's Lady of the Lake can continue to sail on Loch Katrine in her fresh water environment with annual slippage for maintenance is not known. How long the Solent Steam Packet Company can maintain the *Shieldhall* in steam is also uncertain. What is certain is that our regular patronage of these services is essential to help ensure that they remain in steam for the enjoyment of those who would never otherwise have seen a steamship let alone travel in one.

APPENDIX

UK COASTAL STEAMERS AND TUGS REGISTERED IN JUNE 1967 [Ireland had none]

Coastal Steamers

Name	Date built	Grt	Service speed	Comments
Central Electricity Generating Board, London				
Barford	1950	3357	10½	Engines aft collier
Brimsdown	1951	1837	10	Engines aft flatiron collier
Brunswick Wharf	1951	1782	10	Engines aft flatiron collier
Captain J M Donaldson	1951	3357	10	Engines aft collier
Charles H Merz	1955	2947	11	Engines aft collier
Cliff Quay	1950	3357	10½	Engines aft collier
Hackney	1952	1782	10	Engines aft flatiron collier
James Rowan	1955	2947	11	Engines aft collier
Lord Citrine	1950	3357	10½	Engines aft collier
Oliver Bury	1946	2904	10	Engines aft collier
Polden	1950	1362	10	Engines aft flatiron collier
Pompey Light	1949	1428	10	Engines aft flatiron collier
Sir Alexander Kennedy	1946	1714	9	Engines aft flatiron collier
Sir Archibald Page	1950	3357	10	Engines aft collier
Sir John Snell	1955	2947	11	Engines aft collier
Sir Johnstone Wright	1955	2947	11	Engines aft collier
Sir William Walker	1954	3050	11½	Engines aft collier
W J H Wood	1951	3357	10	Engines aft collier
Charrington Gardner Locket, London				
Lady Charrington	1952	2154	10	Engines aft collier
Cory Maritime, London				

*The fully-laden **Brimsdown** heads up the Thames past Woolwich in June 1972.*

(Andrew Wiltshire collection)

*The **Lady Charrington** in also seen in the River Thames.*

(World Ship Photo Library, George Gould collection)

Cormain	1942	2883	10	Engines aft collier
ex *Coldridge*-49; *Cormain*-46				
Cormist	1946	2885	10	Engines aft collier
Corstream	1955	3375	10	Engines aft collier

Ellerman's Wilson Line, Hull

Ariosto	1946	2468	13½	Bauer Wach engines, 12 passengers
Borodino	1950	3206	13½	56 passengers
Cicero	1954	2497	13½	Bauer Wach engines, 12 passengers
Leo	1947	1792	13½	Bauer Wach engines, 12 passengers
Livorno	1946	2957	13	
Rinaldo	1946	2957	13	
Rollo	1954	2499	13	Bauer Wach engines, 12 passengers
Teano	1955	1580	13	Bauer Wach engines
Tinto	1947	1795	13½	Bauer Wach engines, 12 passengers
Truro	1947	1795	13½	Bauer Wach engines, 12 passengers
Volo	1946	1797	13½	Bauer Wach engines, 12 passengers

Esso Petroleum Company Ltd, London

Esso Chelsea	1945	4352	7½	Twin screw tanker
ex *Amacuro*-56				
Esso Fulham	1945	4352	7½	Twin screw tanker
ex *Trujillo*-56				
Esso Preston	1956	2500	10½	Tanker

F T Everard & Sons Ltd, London

Alchymist	1945	813	9	Tanker
ex *Empire Orkney*-50				
Argosity	1941	877	9½	Tanker
ex *Esso Juniata*-56, *Empire Lass*-46				
Aureity	1942	813	9	Tanker
ex *Mascara*-51, ex-*Empire Cadet*-46				
Candourity	1946	474	8½	Tanker
ex *C641*-56				
City	1945	500	7	Tanker
ex *C633*-56				

*A splendid photograph of the **City** arriving at Cardiff on the morning of 20 April 1969.*

(John Wiltshire)

Clanity ex *C642*-56	1946	495	8	Tanker
Commodity ex *C614*-56	1943	469	8	Tanker
Conformity ex *C85*-56	1940	484	9	Tanker
Tankity ex *MOB 7*-57	1945	145	6	Tanker
Totality ex *MOB 13*-57	1946	145	6	Tanker

William France Fenwick & Company Ltd, London

Bearwood	1955	3393	$10^{1}/_{2}$	Engines aft collier
Helmwood	1956	3403	$10^{1}/_{2}$	Engines aft collier

Hudson Steamship Company Ltd, London

Hudson River	1949	3128	$10^{1}/_{2}$	Engines aft collier

Irish Shipping Company, Dublin

Irish Holly	1954	2940	$11^{1}/_{2}$	Tanker

Jersey Shipping Company, St Helier

La Duchesse de Normandie ex *Sir Richard Grenville*-63	1931	913	14	Twin screw, 550 passengers

John Kelly Ltd, Belfast

Ballyhaft	1952	991	10	Engines aft collier
Ballyhill	1954	991	10	Engines aft collier
Ballylagan	1955	1307	10	Engines aft collier
Ballylumford	1954	1242	10	Engines aft collier
Ballymena	1954	1356	10	Engines aft collier
Ballymoney	1953	1342	10	Engines aft collier

North Eastern Fisheries, Aberdeen

Mount Battock	1939	396	$9^{1}/_{2}$	Coaster carrying mainly bunker coal

North Thames Gas Board, London

David Pollock	1954	3332	$10^{1}/_{2}$	Engines aft collier
John Orwell Philips	1955	3378	$10^{1}/_{2}$	Engines aft collier
Sir David II	1954	3332	$10^{1}/_{2}$	Engines aft collier
Thomas Goulden	1955	3332	$10^{1}/_{2}$	Engines aft collier

The John Orwell Phillips passes Tynemouth on her way to load a cargo of coal.

(Stuart Emery collection)

Orkney Islands Shipping Company, Kirkwall

Earl Sigurd	1931	221	9½	50 passengers

C Rowbotham & Sons (Management) Ltd, London

Pointsman	1934	1174	7	Tanker
ex *Bassethound*-59				

Shell-Mex & BP, London

Pando	1921	313	7	Tanker
Perso	1921	313	7	Tanker
Phero	1921	325	7	Tanker
Philo	1921	338	7	Tanker
Poilo	1921	307	7	Tanker

South Eastern Gas Board

Effra	1946	2710	10	Engines aft collier

Stephenson Clarke Ltd, London

Arundel	1956	3422	10½	Engines aft collier
Borde	1953	3401	10½	Engines aft collier
Bramber	1954	1968	9½	Engines aft collier
ex *Greenbatt*-60				
Heyshott	1949	2918	10	Engines aft collier
ex *Colville*-49				

The *Heyshott* in the River Thames. (Stuart Emery collection)

G W Thacker, Newcastle

Kingham	1949	2002		Panamanian flag
ex *Arö*-65				

United Baltic Corporation, London

Baltrover	1949	2179	14	4 passengers
ex *Marstenen*-50				

Steam tugs – [excluding Ministry of Defence] greater than 100 tons gross

Name	Date built	Grt	Indicated horse power	Comments

W H J Alexander (Sun Tugs) Ltd, Thames and Medway

Sun VIII	1919	196	750	
Sun X	1920	196	750	
Sun XII	1925	183	750	
Sun XV	1925	183	750	
Sun XVII	1946	233	1030	

The *Sun XV* at work in the River Thames in 1968.

(Andrew Wiltshire collection)

Alexandra Towing Company Ltd, Mersey, Southampton and Swansea, including Liverpool Screw Towing Company and North West Tugs Ltd

Alfred	1937	215	1000	
Black Cock	1939	168	1000	
Brambles	1942	242	1100	
ex *Empire Teak*-47				
Canada	1951	237	1200	
Canning	1954	200	950	
Caswell	1943	276	850	
ex *Assistant* -62, *Empire Sybil*-47				
Crosby	1937	215	1000	
Fighting Cock	1953	218	1250	
Flying Breeze	1938	460	1000	Tug/tender, 250 passengers
ex *BP Protector*-62, *Zurmand*-55				
Flying Kestrel	1943	244	1000	
ex *Metinda*-49, *Empire Mascot*-47				
Formby	1951	237	1200	
Game Cock V	1953	218	1250	
Gladstone	1951	237	1200	
Grebe Cock	1935	169	1000	
Holm Cock	1934	167	1000	

Marsh Cock	1936	201	1000
ex *Wapping*-67			
North Beach	1956	220	1000
North Buoy	1959	219	1000
North Cock	1936	201	1000
ex *Hornby*-67			
North End	1957	215	1000
North Light	1956	206	1000
North Quay	1956	219	1000
North Rock	1956	206	1000
North Wall	1959	219	1000
Storm Cock	1936	169	1000
Thistle Cock	1929	169	1000
Wallasey	1954	200	950
Waterloo	1954	200	950

Ardossan Harbour Company, Ardrossan

Seaway	1942	260	1000
ex *Empire Palm*-48			

Blyth Harbour Commissioners, Blyth

Chipchase	1953	106	400	Tug/tender, twin screw; 50 passengers

The *Chipchase* at Seaham harbour in September 1972

(Andrew Wiltshire collection)

Blyth Tug Company, Blyth

Francis Batey	1914	151	750
Hillsider	1924	177	800
ex *Headman*-62			
Homer	1915	157	550
Seasider	1919	15	500
ex *West Hyde*-48			

Chas Brand, Belfast

Lavina	1928	300	900
ex *Sloyne*-66			
Lilias	1928	260	950
ex *James Lamey*-66, *Flying Eagle*-59			

British Transport Docks Board, Barrow and Fleetwood

Clevelys	1929	110	400	Twin screw
Rampside	1941	260	850	
ex *Central No 3*-61, *Empire Fir*-46				
Ramsden	1934	188	680	Formerly used as a tender
Roa	1944	232	900	
ex *Central No 4*-61, *Empire Polly*-47				

*The **Ramsden** is almost hidden by Fisher's steam tug **Fishershill** at Heysham on 26 June 1967.*
(John Wiltshire)

Clyde Shipping Company Ltd, Clyde

Flying Merlin	1951	261	1000

John Cooper (Belfast) Ltd, Belfast

Meadow	1942	242	1000
ex *Empire Meadow*-47			
Piper	1942	250	1000
ex *Empire Piper*-47			
Southampton	1910	227	1000

Cork Harbour Commissioners, Cork

Francis Hallinan	1946	296	1000
ex *Foremost 105*-49, *Empire Greta*-47			

Dublin Port & Docks Board, Dublin

Ben Eader	1932	228	1000
ex *Foremost 84*-33			
Coliemore	1926	244	850
ex *Foremost 42*-33			

Trustees of the Harbour of Dundee, Dundee

Harecraig II	1951	261	1180
ex *Flying Buzzard*-63			

Ellerman's Wilson Line, Hull

Forto	1939	180	110
Presto	1943	276	1000
ex *Empire Sara*-46			

Falmouth Towage Company Ltd, Falmouth

Lynch	1924	211	650
ex *Foremost-23*			
St Agnes	1925	226	680
ex *Fairnilee*-59, *Nobleman*-37			
St Denys	1929	174	790
ex *Northgate Scot*-59			
St Levan	1942	160	700
ex *Codicote Scot*-59, *Bruno Dreyer*-51			
St Mawes	1951	346	800
ex *Arusha*-59			
St Merryn	1945	233	1000
ex *Rockpigeon*			

The *St Merryn* in Falmouth Docks on 8 July 1973.

(John Wiltshire)

James Fisher & Sons Ltd, Heysham and Barrow

Fishershill	1946	292	850
ex *Empire Hilda*-49			
Fisherstown	1944	232	850
ex *Empire Roger*-44			

Fowey Harbour Commissioners, Fowey

St Canute	1931	151	500
ex *Othonia*-62, *Sct Knud*-59			

France Fenwick Tyne & Wear Company Ltd, Newcastle and Middlesbrough

Cleadon	1899	148	700
ex *Toxteth*			
Robert Redhead	1900	178	1000
ex *Hannah Joliffe*-30			
Wearmouth	1929	182	900

Gaselee & Son, Felixstowe

Ocean Cock	1932	182	1000

*The **Ocean Cock** approaches Felixstowe Dock on 15 August 1966.*

(John Wiltshire)

Grangemouth & Forth Towing Company Ltd, Grangemouth

Dundas	1931	150	1000
ex **Stronghold**-49			
Kerse	1923	214	800

C J King & Sons Ltd, Avonmouth, Bristol, Portishead

Sea Alarm	1941	263	1000
ex **Flying Fulmar**-56, **Empire Ash**-46			
Sea Queen	1944	244	1000
ex **Empire Walter**-47			

Lawson-Batey Tugs Ltd, Newcastle

Beamish	1944	242	1000	
ex **Queensgarth**-49, **Empire Paul**-46				
Cullercoats	1898	181	700	Oldest tug listed
ex-**Cyclop**				
Eastsider	1924	175	600	
ex **Pluto**-53				
Tynesider	1942	262	1000	
ex **Empire Cherub**-46				

Leith Dock Commissioners, Leith

Mickry	1920	172	750
ex **Vanguard**-46			
Oxcar	1919	252	750
ex **Holland**-26			

London Dredging Company Ltd, Thames

Danube IV	1927	266	829
Danube V	1935	241	900
Danube VI	1935	241	900
Danube VII	1946	237	900
Danube VIII	1946	237	900

Manchester Ship Canal Company, Manchester

Daniel Adamson	1903	175	580	Tug/tender, twin screw
ex **Ralph Brocklebank**-36				
MSC Archer	1938	144	750	
Stanlow	1924	100	480	

88

*The **Dunhawk** is in the lock at Newport on 18 June 1967.*

(John Wiltshire)

Newport Screw Towing Company Ltd, Newport

Dunfalcon   1941   252   1000
    ex **Battleaxe**-61, **Vanguard**-61, **Empire Pine**-48
Dunhawk   1943   244   1000
    ex **Flying Typhoon**-61, **Empire Maisie**-47

J H Pigott & Son Ltd, Immingham and Grimsby

Lady Vera   1938   230   850
    ex **Brahman**-62

Port of London Authority, Thames

Thorney   1943   137   500
    ex **Empire Percy**-48
Westbourne   1912   185   575

R & J H Rea and Rea Towing Company, Liverpool & Bristol Channel

Applegarth   1951   231   1120
Aysgarth   1950   231   1120
Bangarth   1951   231   1120
Grassgarth   1953   231   1120
Rosegarth   1954   231   1120
Throstlegarth   1954   231   1120

W J Reynolds Ltd, Plymouth and Devonport

Antony   1921   137   585
    ex **Corgarth**
Carbeile   1929   110   350
    ex **George Livesey**
Tactful   1928   124   400
    ex **Talbenny**-65, **F T Everard**-51
Trevol   1921   137   650
    ex **Reagarth**

*The **Applegarth** awaits her next trun of duty in Liverpool's Hornby Dock on a sunny August day in 1969.*

(John Wiltshire)

Seaham Harbour Dock Company, Seaham Harbour

Wonder	1921	169	620

ex **Bacalau**-35

Also the paddle tugs **Eppleton Hall** (1914) and **Reliant** (1907) ex **Old Trafford**-50

Southampton, Isle of Wight & South of England Royal Mail Steam Packet Company Ltd, Southampton

Hamtun	1953	318	1500	Twin screw
Sir Bevois	1953	318	1500	Twin screw

Tees Conservancy Commissioners, Tees

Francis Samuelson	1924	140	400

Also the paddle tug **John H Amos** (1931)

United Towing Company Ltd, Humber

Airman	1945	333	750
ex **Empire Clara**-47			
Guardsman	1946	329	750
ex **Empire Nina**-47			
Rifleman	1945	333	750
ex **Empire Vera**-47			

William Watkins Ltd (Ship Towage [London] Ltd), Thames area

Atlantic Cock	1932	182	1000
Cervia	1946	233	900
ex **Empire Raymond**-47			
Challenge	1931	212	1150
Contest	1933	213	1150
Crested Cock	1935	177	1000
Napia	1943	261	1200
ex **Empire Jester**-46			
Tanga	1931	203	850

James A White & Company Ltd, St Davids Harbour

Recovery	1899	194	384	Salvage tug

The Workington Harbour & Dock Company Ltd

Solway	1943	232	900
ex **Empire Ann**-47			

References

Periodicals

Sea Breezes, *Ships Monthly*, *Journal of Commerce*.

Books

Bowen F 1938. *London ship types*. The East Ham Echo Ltd, London.
Brodie I 1976. *Steamers of the Forth*. David & Charles, Newton Abbot.
Chappell C 1980. *Island lifeline*. T Stephenson & Sons, Prescot.
Cowsill M 1990. *By road across the sea, the history of the Atlantic Steam Navigation Company Ltd*. Ferry Publications, Kilgetty.
Duckworth C L D & Langmuir G E 1956. *West coast steamers*. 2nd edition.
Duckworth C L D & Langmuir G E 1967. *West Highland steamers*. 3rd Edition. T Stephenson & Sons, Prescot.
Duckworth C L D & Langmuir G E 1967. *Railway and other steamers*. 2nd Edition. T Stephenson & Sons, Prescot.
Duckworth C L D & Langmuir G E 1972. *Clyde River and other steamers*. 3rd Edition. Brown, Son and Ferguson, Glasgow.
Duckworth C L D & Langmuir G E 1977. *Clyde and other coastal steamers*. 2nd Edition. T Stephenson & Sons, Prescot.
Farr G 1967. *West country passenger steamers*. 2nd Edition. T Stephenson & Sons, Prescot.
Goodwyn A M 1985. *Eight decades of Heysham Douglas*. Manx Electric Railway Society, Douglas.
Hague A 1998. *Convoy rescue ships, a history of the Rescue Service its ships and their crews 1940-1945*. The World Ship Society, Gravesend.
Harrower J 1998. *Wilson Line, the history and fleet of Thos. Wilson, Sons & Co. and Ellerman's Wilson Line Ltd*. The World Ship Society, Gravesend.
Harvey W J and Telford P J 2002. *The Clyde Shipping Company 1815-2000*. P J Telford, UK.
Jordan S 1998. *Ferry services of the London, Brighton & South Coast Railway*. The Oakwood Press, Usk.
Le Scelleur K 1985. *Channel Islands Railway Steamers*. Patrick Stephens, Wellingborough.
Maund T B 1991. *The Mersey Ferries Volume I*. Transport Publishing Company, Glossop.
McNeill D B 1969. *Irish passenger steamship services Volume 1, North of Ireland*. David & Charles, Newton Abbot.
McNeill D B 1971. *Irish passenger steamship services Volume 2, South of Ireland*. David & Charles, Newton Abbot.
Mitchell W H & Sawyer L A 1965. *Empire ships of World War II*. The Journal of Commerce and Shipping Telegraph Ltd, Liverpool.
Paget-Tomlinson E 1980. *North west steamships*. Countryside Publications Ltd., Chorley.
Ridley Chesterton D & Fenton R S 1984. *Gas and electricity colliers, the sea-going ships owned by the gas and electricity industries*. The World Ship Society, Kendal.
Robins N S 1998. *The British excursion ship*. Brown, Son and Ferguson, Glasgow.
Robins N S 1999. *Turbine steamers of the British Isles*. Colourpoint Books, Newtownards.
Robins N S 2003. *Ferry powerful, a history of the modern British diesel ferry*. Bernard McCall, Portishead.
Sinclair R C 1990. *Across the Irish Sea, Belfast-Liverpool shipping since 1819*. Conway Maritime Press, London.
Thornley F C 1962. *Past and present steamers of North Wales*. 2nd Edition. T Stephenson & Sons, Prescot.

*Full steam ahead for the **Danube V** in the lower reaches of the River Thames.*

(World Ship Photo Library, Stuart Emery collection)

MORE STEAM TUGS . . .

All three tugs on this page have been the subject of preservation attempts. The **Canning** is currently on display at the Maritime Museum in Swansea. She was photographed as she emerged from the lock at Swansea on 1 September 1970.

(John Wiltshire)

A colourful view of the **Cervia** leaving the London dock system and entering the Thames in January 1970. The tug is preserved at Ramsgate although sadly her condition is being allowed to deteriorate.

(Andrew Wiltshire collection)

Sadly, the preservation of the **Sea Alarm** at a location adjacent to the former industrial and maritime museum in Cardiff was less than successful. Neglected and deteriorating during the 1980s and 1990s, she was later demolished as part of a redevelopment scheme for the area. We see her here leaving Avonmouth on 30 October 1971.

(John Wiltshire)

. . . AND STEAM COASTERS

The **Craigavad** prepares to depart from Preston on 24 April 1965.

(Jim McFaul)

Another view at Preston, this time finding John Kelly's **Ballyhill** manoeuvring in the port on 20 August 1964.

(Jim McFaul)

The **Sir Archibald Page** outward bound in the Thames in June 1972, no doubt heading for the north-east of England to load a further cargo of coal.

(Andrew Wiltshire collection)

Index

Ship names in ordinary text are British and Irish steam reciprocating engined screw-driven steamers. Ship names in italics denote ships of other nationalities, and diesel, paddle wheel or steam turbine vessels of any nationality. Where known, the year of build is given in brackets.

Name	Page
Aaro (1909)	24
A B Gowan	71
Abercraig (1939)	74
Aberdonian (1909)	35,36
Aberdonian Coast (1947)	36
Abington (1921)	64
Aboyne (1937)	38,43
Acclivity (1929)	66
Accrington (1910)	26,38
Aire (1931)	42
Akabo (1912)	25
Alberta (1900)	18,23
Alchymist (1945)	66
Alma (1894)	18,22
Alnwick (1929)	36
Alsatia (1923)	58
Amacuro (1945)	67
America (1891)	56
Amsterdam (1894)	7,18,19
An Saorstat (1900)	55
Anglia (1900)	11,12,18
Annan (1907)	33
Antrim (1904)	14,18
Aranmore (1920)	46
Ardetta (1949)	43
Ardmore (1921)	28,31
Ardri (1892)	65
Ardyne (1928)	42
Argosity (1941)	66
Ariosto (1940)	44
Ariosto (1946)	44
Artificer (1905)	41
Arundel (1900)	18,20,21
Arundel (1956)	62,63,68
Arvonia (1897)	12
Atalanta (1906)	59
Atalanta (1907)	59
Atheltarn (1929)	66
Atlantic Coast (1934)	32
Aureity (1942)	66
Autocarrier (1931)	41,42
Averity (1944)	66
Avon (1887)	35
Baldur (1937)	51
Bali (1929)	36
Ballybeg (1898)	63
Ballyhill (1954)	63
Ballygarvey (1937)	63
Ballylagan (1955)	62,63
Ballylumford (1954)	63
Ballymena (1954)	63
Ballymoney (1953)	63
Baltabor (1911)	25
Baltallin (1920)	25
Baltannic (1916)	25
Balteako (1920)	25
Baltonia (1912)	25
Baltriga (1916)	25
Baltrover (1913)	25
Baltrover (1949)	25
Bardic Ferry (1957)	6
Barnston (1921)	69
Bayardo (1911)	24
Beachy (1936)	33,38
Beauly (1924)	33
Bebington (1925)	69
Ben Maye (1921)	64
Ben-my-Chree (1845)	10
Ben-my-Chree (1875)	10
Ben-my-Chree (1908)	10
Ben-my-Chree (1966)	76
Berlin (1894)	18,19
Bernicia (1923)	36
Bidston (1903)	56
Bidston (1933)	55,70
Bison (1906)	58
Bittern (1949)	43
Blackburn (1910)	26
Blarney (1906)	55,71
Blyth (1931)	42
Bolton Abbey (1957)	26
Borde (1953)	65,68
Borodino (1950)	6,45
BP Protector (1938)	58
Braedale (1894)	65
Bramber (1954)	68
Bramhill (1923)	65
Brier (1882)	26,27
British Coast (1933)	32
Brora (1924)	33
Brussels (1902)	25
Bury (1910)	26,38
Calabar (1936)	5
Calder (1931)	42
Caledonian Coast (1948)	36
Calshot (1930)	57,58,61,74
Calypso (1904)	24
Cambria (1897)	11,12,18
Cambria (1949)	74
Captain J M Donaldson (1951)	63
Caripito (1945)	67
Carlo (1947)	44
Carlotta (1893)	72
Carron (1909)	35
Carrowdore (1914)	66
Catherine (1903)	69,72
Celtic (1901)	15
Chant 53 (1944)	66
Charles Galloway (1885)	57
Charles H Merz (1955)	63
Chelmsford (1893)	19
Chieftain (1907)	48
Chieftain (1930)	76
Chipchase (1956)	58
Churton (1921)	69
Cicero (1954)	45
City of Amsterdam (1921)	43
City of Antwerp (1915)	43
City of Belfast (1893)	18
City of Cork (1917)	43
City of Hamburg (1937)	43
City of Waterford (1921)	43
Clangula (1954)	42,43
Clarecastle (1914)	66
Clareisland (1914)	66
Classic (1893)	15
Claughton (1930)	70
Claymore (1881)	48
Cliff Quay (1950)	63
Columbia (1894)	18,22
Conister (1921)	11,64
Connaught (1860)	11
Connaught (1897)	11,12,18
Connemara (1897)	12,13,18
Consuelo (1937)	44
Copeland (1923)	33,34,37,38,40
Copenhagen (1907)	19
Cork (1899)	30
Countess of Bantry (1884)	51
Courland (1932)	43
Craigantlet (1931)	65
Craigavad (1924)	65
Craigolive (1921)	65
Daffodil (1906)	55,70
Dafila (1927)	43
Dalmarnock (1925)	77
Dalmuir (1904)	77
Dalriada (1920)	47
Daniel Adamson (1903)	57
Davaar (1885)	47
Deal (1925)	41
Dearne (1925)	41
Deerhound (1901)	54
Delgany (1921)	43
Devonshire (1934)	58
Dewsbury (1910)	26,38
Digby (1913)	25
Dinard (1924)	42
Divis (1928)	77
Dominica (1913)	25
Don (1924)	41
Donaghadee (1937)	63
Donegal (1904)	14,18
Dorothy Hough (1911)	32
Dotterel (1936)	43
Douglas (1858)	10
Douglas (1864)	10
Dresden (1896)	18,19
Drover (1923)	33
Duchess of Devonshire (1897)	18
Duke of Albany (1907)	14,18
Duke of Argyll (1909)	14
Duke of Clarence (1892)	13,18,20
Duke of Connaught (1902)	14,18,20
Duke of Cornwall (1898)	13,18
Duke of Cumberland (1909)	14
Duke of Lancaster (1895)	8,13,18
Duke of Montrose (1906)	27
Duke of Rothesay (1899)	27,29
Duke of York (1894)	13,18
Dunara Castle (1875)	48
Dundee (1934)	38
Dunure (1878)	26
Earl of Zetland (1877)	51
Earl of Zetland (1939)	51
Earl Sigurd (1931)	4,51
Earl Thorfinn (1928)	51
Eddystone (1927)	38
Edith (1911)	69,72
Edward Cruse (1954)	63,77
Egerton (1911)	56,57
Egret (1936)	43
Eldorado (1886)	24
Ellan Vannin (1886)	5,10
Empire Baltic (1945)	45
Empire Cadet (1942)	66
Empire Cedric (1945)	45
Empire Celtic (1945)	45
Empire Lass (1941)	66
Empire Nordic (1945)	45
Empire Orkney (1945)	66
Empire Peggotty (1944)	66
Empire Shearwater (1945)	45
Empress (1907)	19
Empress Queen (1897)	10
Endrick (1928)	43
Erskine (1925)	72
Eskimo (1910)	24
Esso Chelsea (1945)	66,67
Esso Fulham (1945)	66,67
Esso Lambeth (1945)	66
Esso Preston (1956)	62,63,64
Esso Wandsworth (1945)	66,67
Failte (1901)	55
Farringford (1947)	8
Fastnet (1928)	38
Fenella (1881)	5,10
Ferry Princess (1959)	71
Findhorn (1903)	33

Flying Breeze (1913)	56,57,58	King Orry (1913)	10	Marstenen (1949)	25
Flying Breeze (1938)	58	Kyle Rhea (1921)	64	Marylebone (1906)	7
Flying Buzzard (1951)	58	La Brétonnière (1907)	59	Mecklenburg (1922)	6,19
Flying Childers (1856)	46	La Duchesse de Normadie (1931)	60	Megstone (1946)	63
Flying Eagle (1880)	46	Lady Carlow (1896)	30	Mellite (1873)	67
Flying Kestrel (1913)	58	Lady Cloé (1915)	32	Melrose Abbey (1929)	26,38
Flying Merlin (1951)	58	Lady Connaught (1906)	16	Melrose Abbey (1959)	26
Forde (1919)	41,42	Lady Connaught (1911)	16	Menevia (1902)	12
Francis Storey (1922)	55,70	Lady Gwendolen (1911)	32	Mimie (1927)	72
Fratton (1925)	41	Lady Kerry (1897)	30	Minard (1926)	42
Frederica (1890)	18,21,22,23	Lady Killarncy (1911)	16	Minden (1903)	55,56
Frinton (1903)	30	Lady Leinster (1911)	16	Miranda Guinness (1976)	66
Furness (1934)	58	Lady Limerick (1924)	28	Mona (1832)	10
Galtee More (1898)	12,13,18	Lady Longford (1921)	28,31	Mona (1878)	5,10
Galway Bay (1930)	33,58,61	Lady Louth (1894)	28,30	Mona (1907)	16
Galway Coast (1915)	33	Lady Martin (1888)	32	Mona's Isle (1830)	10
Gertrude (1908)	55,69,72	Lady Munster (1906)	16	Mona's Isle (1860)	5,10
Glanowen (1944)	66	Lady Olive (1878)	32	Mona's Isle (1882)	5,10
Glenageary (1920)	65	Lady Roberts (1897)	32	Mona's Queen (1852)	10
Glencree (1934)	65	Lady Savile (1891)	59,60	Mona's Queen (1885)	10
Glencullen (1921)	65	Lady Wicklow (1895)	30	Moorfowl (1919)	26,27
Glengariff (1936)	6,31,33	Lady Wimborne (1915)	32,33	Morsecock (1877)	58
Goodwin (1917)	34,38,43	Lairds Loch (1944)	28	Mount Etna (1880)	48
Gothland (1932)	38,43	Lairdsbank (1893)	27,28,50	Munster (1860)	11
Graphic (1906)	14,15,16,18	Lairdsburn (1923)	28,56	Munster (1897)	11,12,18
Great Southern (1902)	31	Lairdscastle (1924)	28	Nomadic (1911)	58
Great Western (1902)	31	Lairdscraig (1936)	33	North Buoy (1959)	76
Great Western (1934)	6	Lairdsford (1899)	27,29	North Down (1923)	33
Guarico (1945)	67	Lairdsforest (1906)	27	North Tipperary (1917)	43
Guinness (1931)	65,66	Lairdsglen (1914)	27,28,30	North Wall (1959)	76
Gwentland (1923)	65	Lairdsgrove (1898)	27,28,30	Northumbrian (1930)	69,71
Hadrian (1923)	36	Lairdshill (1921)	28,31	Olive (1893)	26,27,28,50
Halcyon (1921)	38,44	Lairdsloch (1906)	27	Oranje Nassau (1909)	19
Halladale (1946)	42	Lairdsmoor (1919)	27	Orcadia (1963)	51
Harecraig II (1951)	58	Lairdsoak (1882)	27	Orlando (1904)	25
Haringvliet (1932)	71	Lairdspool (1896)	27	Oslo (1906)	24
Harrogate (1925)	41	Lairdsrock (1898)	27	Oxton (1925)	69
Haslemere (1925)	41	Lairdsrose (1902)	27,28,29,30	Paladin (1913)	56,58
Hazel (1907)	16,18	Lairdsvale (1896)	27	Partridge (1906)	26,27,30
Hebble (1924)	41	Lairdswood (1906)	27	Pass of Balmaha (1933)	66
Hebrides (1898)	48	Laverock (1909)	44	Patriotic (1911)	14,15,16,18
Helmwood (1956)	62	Leasowe (1921)	69	Peel Castle (1894)	13
Hengist (1928)	43	Leinster (1860)	11	Peninnis (1904)	52
Heroic (1906)	14,15,16,18	Leinster (1897)	11,12,17,18	Peninnis (1925)	52
Heron (1920)	25,44	Leo (1947)	44	Penlee (1891)	59
Heyshott (1949)	68	Lily (1896)	26,27	Perch Rock (1929)	69
Hibernia (1900)	7,11,12,18	Lily (1901)	55	Perth (1915)	35,38
Hibernian Coast (1947)	36	Liscard (1921)	69	Peveril (1884)	5,10
Highland (1951)	63	Lochbroom (1945)	48	Peveril (1929)	52
Hinderton (1925)	70	Loch Frisa (1946)	48	Philomel (1921)	38,44
Holdernith (1944)	66	Lochfyne (1931)	73	Philomel (1927)	44
Hörnum (1919)	54	Lochgarry (1898)	27	Poilo (1921)	66
Horsa (1928)	43	Lochnagar (1906)	35	Pointer (1896)	26,27
Hound (1893)	26	Lochness (1930)	48	Prenton (1906)	69
Hythe (1925)	41	London	35	Prince of Wales (1887)	10
Ibex (1891)	18,22	Londonderry (1904)	14	Prinses Juliana (1920)	19
Icemaid (1936)	66	Longford (1906)	16	Puma (1899)	26,27
Immingham (1906)	7	Longships (1917)	34	Queen (1861)	53
Ingénieur Reibell (1908)	60	Lord Stalbridge (1909)	57	Queen of the Isle (1834)	10
Ingénieur Reibell (1911)	60	Lorrain (1908)	13	Queen Victoria (1887)	10
Inniscarra (1903)	31	Lotharingia (1923)	58	Rajula (1926)	5,6
Innisfallen (1896)	31	Louth (1906)	16,30	Ralph Brocklebank (1903)	57
Ionic Ferry (1958)	6	LST 3424 (1945)	45	Ramsden (1934)	58,59
Ireland (1891)	56	LST 3512 (1945)	45	Ramsey Town (1904)	14
Iris (1906)	55,70	LST 3519 (1945)	45	Rapallo (1960)	75
Irish Coast (1952)	16	Lydia (1890)	18,21,23	Rathlin (1936)	33,38,46
Irish Holly (1954)	62	Lymington (1939)	74	Rathmore (1908)	12,13,18
J Farley (1922)	70	Lyon (1885)	20	Reindeer (1897)	18,21,22
James Rowan (1955)	63	Lyonesse (1889)	5,52	Renfrew (1925)	72
John Joyce (1910)	55	Magic (1893)	14,15,18	Richard Welford (1908)	36
Joseph Swan (1938)	62	Magic II (1893)	15	Ringwood (1926)	41
Kenmare (1921)	28,31,33	Magnetic (1891)	58	Robina (1914)	54
Kilkenny (1903)	30	Magpie (1898)	26,27,28,30	Rochester Queen (1908)	55,72
Killarney (1893)	15	Malmo (1947)	44	Rockabill (1931)	40,46
Killarney (1922)	55,70	Mancunium (1933)	77	Roebuck (1897)	18,21,22
King Edward (1901)	7,54	Mancunium (1946)	77	Roebuck (1925)	40
King Harry Ferry	72	Manxman (1904)	14	Rollo (1899)	25
King Orry (1842)	10	Maple (1914)	26,27,30	Rollo (1954)	45
King Orry (1871)	10	Marlowe (1927)	70,71	Romeo (1881)	24

Romsey (1930)	56,58	Selby (1922)	41	Toward (1923)	33,37,38,46
Rose (1900)	69,72	Shandon (1910)	55	*Traffic* (1911)	58,60
Rose (1901)	55	Sheila (1904)	48	*Trujillo* (1945)	67
Rose (1902)	26,27,29,30	Shieldhall (1910)	77	Truro (1947)	44
Roslin Castle (1906)	54	Shieldhall (1955)	4,63,76,77,78,79	Tuskar (1920)	40,46
Rosstrevor (1895)	12,13,18	Silvio (1947)	44	Tynesider (1925)	69,71
Roulers (1894)	19	Sir Francis Drake (1908)	59	*Tynwald* (1846)	10
Royal Archer (1928)	35	Sir John Hawkins (1929)	59,60	*Tynwald* (1866)	10
Royal Daffodil (1906)	55,70	Sir John Snell (1955)	63	Tynwald (1891)	5,10
Royal Daffodil II (1934)	70,71	Sir Johnstone Wright (1955)	63	*Ulster* (1860)	11
Royal Firth (1921)	41	Sir Richard Grenville (1891)	59	Ulster (1897)	11,12,17,18
Royal Fusilier (1924)	35	Sir Richard Grenville (1931)	59,60	Ulster Duke (1906)	16
Royal Highlander (1910)	35	Sir Walter Raleigh (1908)	59	Ulster Herdsman (1923)	6,33,34,37
Royal Iris (1906)	70	Sir Walter Scott (1900)	4,11,79	*Ulster Monarch* (1929)	8,11,16,73
Royal Iris (1932)	71	Sir William Walker (1954)	63	Umtali (1936)	5
Royal Iris (1950)	71	Skerries (1921)	40,43,46	Umtata (1938)	5
Royal Iris II (1932)	70,71	Skirmisher (1884)	58	Upton (1925)	56,70
Royal Scot (1910)	35	Slemish (1923)	65	*Vecta* (1938)	74
Rushen Castle (1898)	13	Slieve Donard (1922)	39	Vehicular Ferry Boat No 1 (1890)	72
Ryde (1891)	58	Slievebloom (1908)	39	Vehicular Ferry Boat No 2 (1900)	72
Rye (1924)	41	Smeaton (1883)	59	Vehicular Ferry Boat No 3 (1908)	72
St Catherine (1893)	50	*Snaefell* (1863)	10	*Velocity* (1821)	53
St Clair (1937)	6,31,38,50	*Snaefell* (1876)	10	Venus (1948)	72
St Clair (1960)	49	Snaefell (1910)	5,10,18,52	Vera (1898)	18,22
St Clair II (1937)	51	Snowdrop (1910)	56	Vesta	71
St Clement (1928)	51	Southern Coast (1911)	32	Vienna (1894)	18,19,20
St Elian (1919)	54	*Sovereign* (1836)	53	*Viking* (1905)	5,10,52
St Hilary (1932)	71	Stanlow (1924)	57	Viper (1906)	27
St Hilary (1934)	71	Starling (1920)	25,44	*Volatile* (1900)	56
St Magnus (1864)	53	Steersman (1936)	74	Volo (1938)	44
St Magnus (1924)	6,49	Stella (1890)	18,21,22,23	Volo (1946)	44
St Magnus (1937)	51	Stockport (1911)	26,38	Vulture (1898)	26,27
St Magnus II (1937)	51	Stormcock (1877)	58	W E Dorrington (1906)	57
St Margaret (1907)	49	Stormlight (1957)	67,76	W M Barclay (1898)	65
St Ola (1892)	51	Sussex (1896)	18,20,21	Wallasey (1881)	69
St Ola (1951)	51	Tadorna (1928)	43	Wallasey (1927)	70,71
St Seiriol (1914)	54	*Talisman* (1935)	8	*Waverley* (1864)	53
St Seiriol (1931)	54	Tasso (1922)	25	*Waverley* (1947)	4,77
St Sunniva (1931)	38,49	Tasso (1938)	44	Whitstable (1925)	41
St Trillo (1936)	54	Tasso (1947)	44	Wicklow (1895)	30
St Tudno (1926)	54	Tessa (1924)	72	Winga (1957)	62
Salford City (1928)	63	Thane of Fife (1910)	56	Winneba (1938)	5
Sambur (1925)	40	The Ramsey (1895)	13	Woodcock (1906)	26,27,30,35
Savannah (1819)	4	The Rose (1909)	57	Wyresdale (1925)	72
Scillonian (1926)	52	Thistle (1884)	26	Wyvern (1905)	57
Scillonian (1956)	52	Thurstaston (1930)	70	*York* (1959)	75
Scotia (1902)	11,12,18	Tiger (1906)	26,27	Yorkshireman (1928)	56
Seacombe (1901)	69	Tinto (1947)	44	Zaafaran (1920)	38,44
Seaford (1894)	18,20,21	*Titanic* (1912)	15	Zamalek (1921)	38,44
Seamew (1915)	43	Tod Head (1921)	64		
Seamore (1891)	56	Tonbridge (1924)	41		

*Liverpool's renowned Pier Head with two Isle of Man Steam Packet Company vessels in view. Nearer the camera is the steamer **Mona** (1907) and astern of her is the turbine steamer **Manxman** (1904).*

(Author's collection)